Graphene Devices

Graphene Devices

Nelson Bolivar

ARCLER
PRESS

www.arclerpress.com

Graphene Devices

Nelson Bolivar

Arcler Press
224 Shoreacres Road
Burlington, ON L7L 2H2
Canada
www.arclerpress.com
Email: orders@arclereducation.com

ISBN: 978-1-77469-050-5 (Hardcover)

ABOUT THE AUTHOR

Nelson Bolivar is currently a Physics Professor in the Physics Department at the Universidad Central de Venezuela, where he has been teaching since 2007. His interests include quantum field theory applied in condensed matter. He obtained his PhD in physics from the Universite de Lorraine (France) in 2014 in a joint PhD with the Universidad Central de Venezuela. His BSc in physics is from the Universidad Central de Venezuela.

TABLE OF CONTENTS

LIST OF FIGURES

LIST OF TABLES

LIST OF ABBREVIATIONS

2D	Two-Dimensional
3D	Three-Dimensional
AFM	Atomic Force Microscopy
AGNRs	Armchair Graphene Nanoribbons
AM	Air Mass
APRES	Angular Distribution Photoemission Spectroscopy
ASC	Adult Stem Cells
ASE	Anomalous Skin Effect
Au	Gold
BJT	Bipolar Junction Transistor
BN	Boron-Nitride
BSA	Bovine Serum Albumin
BTBT	Band-to-Band Tunneling
CB	Carbon Black
CB	Conduction Band
CCG	Chemically Converted Graphene
CMOS	Complementary Metal Oxide Semiconductor
CNT	Carbon Nanotube
CT	Charge Transfer
CV	Cyclic Voltammetry
CVD	Chemical Vapor Deposition
D-A	Donor-Acceptor
DIBL	Drain-Induced-Barrier Lowering
DLC	Dielectric and Diamond-Like Carbon
DMAc	Dimethylacetamide
DMF	Dimethylformamide
DMH	Dimethylhydrazine

DMSO	Dimethyl Sulfoxide
EA	Electro-Absorption
ECM	Extracellular Matrix
EDLC	Electric Double-Layer Capacitance
EDP	Energy-Delay Product
EM	Electromagnetic
ER	Edge Roughness
ESC	Embryonic Stem Cells
FE	Field Emission
FET	Field-Effect Transistor
FF	Fill Factor
FMI	Ferromagnetic Insulator
FTO	Fluorine-Doped Tin Oxide
GFETs	Graphene FETs
GICs	Graphite Intercalation Compounds
GNR	Graphene Nanoribbon
GO	Graphene Oxide
GQDs	Graphene Quantum Dots
HMPA	Hexamethyl Phosphoramide
HOPG	Highly Ordered Pyrolytic Graphite
IC	Integrated Circuit
IgE	Immune Globulin E
IPCE	Incident Photon to Converted Electron
iPS	Induced Pluripotent Stem
ITO	Indium Tin Oxide
ITRS	International Technology Roadmap for Semiconductors
LB	Langmuir-Blodgett
LCD	Liquid Crystal Display
LED	Light-Emitting Diode
MLGNR	Multilayer GNR
MOSFET	Metal-Oxide-Semiconductor Field-Effect Transistor
MR	Magnetoresistance
M-ZGNR	Magnetized ZGNR
NB	Nanobelt

NDR	Negative Differential Resistance
NEGF	Nonequilibrium Green's Function
NHE	Normal Hydrogen Electrode
NMP	N-Methyl Pyrrolidone
NMR	Nuclear Magnetic Resonance
NW	Nanowire
OLED	Organic Light-Emitting Display
P3HT	Poly(3-hexylthiophene)
P3OT	Poly(3-octylthiophene)
PCE	Power Conversion Efficiency
PdNPs	Pd Nanoparticles
PEG	Polyethylene Glycol
PEGylated	Polyethylene Glycosylated
PEI	Poly Ethylene Imine
PET	Polyethylene Terephthalate
PL	Photoluminescence
PMMA	Poly(methyl methyacrylate)
PSA	Prostate-Specific Antigen
Pt	Platinum
PVDF	Polyvinylidene Fluoride
PVR	Peak-to-Valley Ratio
QDs	Quantum Dots
QDSSC	QD-Sensitized Solar Cells
RF	Radio Frequency
rGO	Reduced Graphene Oxide
RMGO	Reduced Multilayer Graphene
RTD	Resonant Tunneling Diode
SAM	Saturable Absorber Mirrors
SB	Schottky Barrier
SERS	Surface-Enhanced Raman Spectroscopy
SiC	Silicon Carbide
SiO_2	Silicon Dioxide
SOC	Spin-Orbit Coupling
SPP	Surface Plasmon Polariton

SPR	Surface Plasma Resonance
SQD	Semiconductor Quantum Dots
SS	Subthreshold Slope
ssDNA	Single-Stranded DNA
STM	Scanning Tunneling Microscopy
SWCNTs	Single-Walled Carbon Nanotubes
TCD	Total Color Difference
TCO	Transparent Conducting Oxide
TEM	Transmission Electron Microscopy
TFETs	Tunneling Field-Effect Transistors
TGA	Thermogravimetric Analysis
THG	Third-Harmonic Generation
TMPyP	Tetrakis (1-Methyl-4-Pyridinio) Porphyrin
TNT	Trinitrotoluene
UHV	Ultrahigh Vacuum
XRD	X-ray Diffraction
ZGNRs	Zigzag Graphene Nanoribbons

PREFACE

Graphene came out as a novel miracle material in the early 2000s. It successfully caught the imagination and attention of scientists as well as the general public. It would have been hard to fathom about the discovery of a material having a single layer of atoms in the past. However, the discovery of graphene marked the beginning of the new era of materials development.

Multiple random attempts carried out to synthesize and study graphene. We can trace back the initial sporadic attempt to investigate graphene to 1859. However, graphene research showed immense expansion after the discovery and isolation of a single atomic layer of carbon atoms (graphene) in 2004 by Andre Geim and Kostya Novoselov at the University of Manchester in the United Kingdom. They received a Nobel Prize in Physics for recognition of their groundbreaking discovery in 2010.

Professor Geim and Professor Novoselov named their graphene synthesis technique the 'Scotch tape method.' This method used at Manchester University was so convenient and efficient that graphene synthesis grew tremendously fast, and now thousands of laboratories around the globe deal with various aspects of graphene investigation. The Scotch tape method is also called the micromechanical cleavage method which does not necessitate complex equipment or huge investments. Low budget requirements and convenient methodology have considerably facilitated the widespread exploration of graphene science.

Efficacious segregation of single-layered graphene has instigated a lot of curiosity in unraveling the prospects of its use in the synthesis of various electronic and electrical devices, predominantly because of its remarkable electronic characteristics. Graphene is predicted to follow Moore's law by replacing silicon in various electronic applications. This book reviews different aspects of graphene and its applications in the synthesis of various electronic devices.

This book consists of eight chapters. Each chapter provides a comprehensive overview of the topic with emphasis on the practical uses of graphene and its derivatives. Chapter 1 discusses the fundamental concepts of graphene which include functionalization, doping, and characterization of graphene materials. Chapter 2 illustrates the information regarding synthesis routes, classifications, and properties of graphene. Chapter 3 focuses on the general applications of graphene-based materials in the electronics industry. Applications of graphene and graphene nanoribbons (GNRs) in different electronic and optical devices are discussed in the chapter.

Chapter 4 specifically focuses on the uses of graphene in the synthesis of electronic sensors which include gas sensors, pH sensors, biosensors, and metal sensors, Chapter 5 deals with the applications of graphene in the synthesis of solar cells. Chapter 6 briefly discusses the photonic applications of graphene. Photonic applications of graphene include photodetectors, polarizers, plasmonics, and optical devices.

Chapter 7 deals with the synthesis of graphene-based nanocomposites and their applications in superconductors. Finally, Chapter 8 discusses the electronic properties of graphene-based materials and their use in the synthesis of electronic transistors.

This book can be used by materials scientists, chemists, and physicists to expand their knowledge about synthesis, characterization, and applications of graphene-based materials. Moreover, this book can be used as a ready reference by students, engineers, teachers, and industrialists from a diverse background.

—Author

CHAPTER 1

Fundamentals of Graphene

CONTENTS

1.1. INTRODUCTION

Carbon is derived from the Latin word carbo that means charcoal. This element makes more recognized stable allotropes than any other element because of its unique electronic structure that lets hybridization develop sp, sp^2, and sp^3 networks. Graphite is the most common allotropic form of carbon, which has been known since ancient times, like a diamond, and it is found naturally in rich quantity. sp^2 hybridized carbon atomic layers attached by weak van der Waals forces compose graphite. Graphene consists of single layers of carbon atoms firmly arranged into a two-dimensional (2D) honeycomb crystal lattice. Its name was given by Boehm et al. (1994). An exceptional anisotropic behavior related to electrical and thermal conductivity is exhibited by the graphite. It is strongly conductive in the way parallel to the graphene layers due to the in-plane metallic character, while poor conductivity is shown in the direction perpendicular to the layers due to the fragile van der Waals interactions between them (Chung, 2002). Three σ bonds are formed by the carbon atoms in the graphene layer with adjacent carbon atoms by coinciding of sp^2 orbitals whereas, the remaining $_{Pz}$orbitals overlap to make a band of filled π orbitals known as the valence band, and a band of empty π* orbitals known as the conduction band (CB), that are capable for the high *in-plane* conductivity (Pop et al., 2012; Chamanara and Caloz, 2015).

The interplanar spacing of graphite comes to 0.34 nm and is not enormous enough to host organic ions/molecules or alternate inorganic species. Although, many intercalation strategies have been practiced for the magnification of the interlayer galleries of graphite from 0.34 nm to higher values, which can arrive in few cases at more than 1 nm based on the size of the guest species. Since the early intercalation of potassium in graphite, there is a testing of a plethora of chemical species for construction that is called graphite intercalation compounds (GICs). The inserted species are balanced within the graphene layers via polar or ionic interactions without manipulating the graphene structure. Such compounds can be produced not only with sodium, potassium, lithium, and other alkali metals, but also with anions like bisulfate, halogens, or nitrate (Perrozzi et al., 2014).

In alternate cases the occurrence of the insertion of guest molecules might be possible from covalent bonding through chemical grafting reactions among the interlayer space of graphite; this prompt structural changes of the graphene planes as the hybridization of the reacting carbon atoms transforms from sp^2 to sp^3. A property instance is the insertion of strong acids and oxidizing reagents that makes oxygen functional groups at the edges of the graphene layers and on the surfaces and increase the graphite oxide. Firstly, Schafheutl (in 1840) and after 19 years Brodie (in 1859) were the pioneers in the formation of graphite oxide (Somani et al., 2006; Obeng and Srinivasan, 2011). The previously formed graphite oxide with a mixture of nitric and sulfuric acid, whereas the recent treated natural graphite with fuming nitric acid and potassium chlorate. A variation of the Brodie method was proposed by Staudenmaier (1898), where graphite is oxidized by the inclusion of concentrated nitric and sulfuric acid with potassium chlorate. A century later (in 1958) it is reported by Offeman and Hummers (1958) that the oxidation of graphite plus the formation of graphite oxide on immersing natural graphite in a mixture of $NaNO_3$, $KMnO_4$, and H_2SO_4 as a consequence of the reaction of the anions introduced within the graphitic layers with carbon atoms, which shatters the aromatic character. The strong oxidative action of these species results in the production of anionic groups on graphitic layers, generally carboxylates, epoxy, and hydroxylates groups. The distance between the graphene layers is increased from 0.35 nm in graphite to almost 0.68 nm in graphite oxide by the out of planar C-O covalent bonds (Bourlinos et al., 2003; Zhou et al., 2011). This gave rise to the spacing and the polar or anionic character of the oxygen groups made impart to graphene oxide (GO) a powerfully hydrophilic behavior, which makes it possible to pass through water molecules between the graphene layers and by that inflate the interlayer distance even more. Hence graphite oxide becomes eminently dispersible in water. During oxidation, the production of sp^3 carbon atoms rattles the delocalized π system, and resulting electrical conductivity in graphite oxide declines reaching within 10^3 and 10^7 Ωcm based on the amount of oxygen (Allen et al., 2010; Liu et al., 2011).

1.2. GRAPHENE AND GRAPHENE OXIDE (GO)

For many decades, it has seemed that the isolation of graphene monolayer is impossible on the basis of theoretical studies on the thermodynamic stability of 2D crystals, and among other things (Prezhdo, 2011). In 2004, a research group in Manchester directed by Novoselov et al. (2004) made an important step in this way. It reported a procedure for the making of single-layer graphene on a silicon oxide substrate by exfoliating the graphite with micromechanical cleavage (scotch tape procedure). Graphene showed outstanding electrical, mechanical, and structural characteristics, and after 6 years, Geim and Novoselov were esteemed with the Nobel Prize in Physics in the subject of "*for groundbreaking experiments regarding the two-dimensional material graphene*" (Meyer et al., 2007; Castro Neto and Guinea, 2009). Over this time, a number of methods have been developed for the formation of graphene monolayers. These methods can be classified into different types based on the physical or chemical process employed to get the single-layer graphene. Coming three sections define the three categories of chemical methods (Frank et al., 2007; Chen et al., 2010).

1.2.1. Preparation of Graphene From Graphene Oxide (GO)

Even though from the processes organized by Hummers and Offeman (1958), the report was obtained on single sheets of GO and had been published, the scientific community widely proceed to assume graphite oxide a layered graphitic material (Boehm et al., 1962; He et al., 1998). It was not until after the micromechanical cleavage isolates the pristine graphene, the question was reconsider, and it was determined that the exfoliated oxidized single graphene layers are produced from the dispersal of graphite oxide in water by the method developed by Offeman and Hummers. These chemically made monolayers of GO can be imagined as the precursors for the formation of graphene by the elimination of the oxygen groups. The precise structure of GO relies on the oxidation process and is still a topic of discussion. The Dek' any models and the Lerf-Klinowski are the most accepted models (Stankovich et al., 2007; Gao et al., 2009). Recently it is confirmed by Stankovich et al.(2006) that for GO prepared with the set of rules that contributed to the Lerf-Klinowski model, ring lactols are located at the edges of the GO sheets (see Figure 1.1).

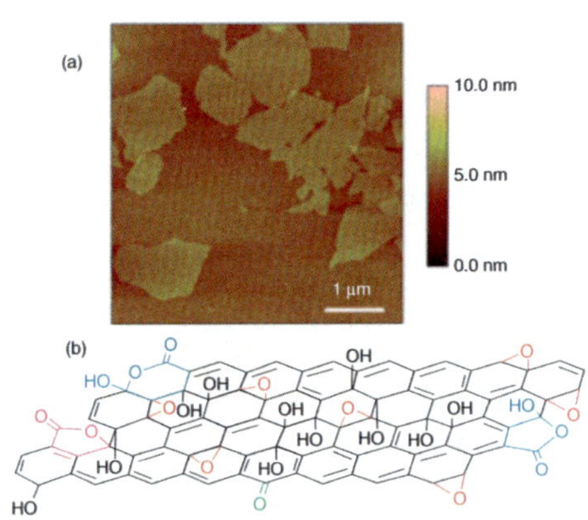

Figure 1.1. Atomic force microscopy (AFM) image and the structural model of graphene oxide (GO) sheets. (a) An AFM image of GO sheets on a silicon substrate; (b) a structural model of GO proposed by Gao et al. (2009).

Source: https://pubmed.ncbi.nlm.nih.gov/21378895/.

In 2006, the first dispersion of single graphene layers was proposed by Ruoff's group, which makes use of hydrazine hydrate prepared by the Hummers method for the reduction of GO (Boukhvalov and Katsnelson, 2008). Although various reductive methods have been introduced by different research groups (refers to Table 1.1 and references within) in the following years, but none of them was able to achieve a full reduction of the GO monolayers into graphene. From the theoretical finding, it is agreed that a reduction of GO from 75% to 6.25% (C:O ratio 16:1) coverage is comparably easy but more reduction seems to be a little difficult. That is why the final isolated carbon monolayers extracted from the reduction of GO are generally known as chemically converted graphene (CCG) or partially reduced graphene oxide (rGO). The consequence of the numerous reductive procedures that have been applied is sum up in the above-given table (Shin et al., 2009; Fernández-Merino et al., 2010).

Table 1.1. Review of Reduction Agents for Chemical Reduction of Graphene Oxide (Mittal, 2012)

Reduction Agent	**Reduction Time (*h*)**	**Temperature (°C) During Reduction**	**Electrical Conductivity (Sm^{-1}) After Reduction**	**References**
Alcohols	24	100	~2.2×10^3	Dreyer et al. (2011)
Alkali	A few minutes	50–90	–	Fan et al. (2008)
Aluminum powder (with hydrochloric acid)	0.5	25	~2.1×10^3	Fan et al. (2010, 2011)
Amino acid (L-cysteine)	12–72	25	–	Chen et al. (2011)
Ascorbic acid (vitamin C)	24	95	~7.7×10^3	Fernández-Merino et al. (2010)
Benzylamine	1.5	90	–	Liu et al. (2011)
Dimethylformamide	1	153	~1.4×10^3	Schniepp et al. (2006)
Hydrazine	24	100	~2×10^2	Boukhvalov and Katsnelson (2008)

Hydroiodic acid	1	100	~3 × 10^4	Pei et al. (2010)
Hydroiodic acid (with acetic acid)	40	40	~3.0 × 10^4	Moon et al. (2010)
Hydroquinone	20	25	–	Dreyer et al. (2011)
Hydroxyl amine	1	90	~1.1 × 10^2	Zhou et al. (2011)
Iron powder (with hydrochloric acid)	6	25	~2.3 × 10^3	Fan et al. (2011)
Pyrogallol	1	95	~4.9 × 10^2	Fernández-Merino et al. (2010)
Sodium borohydride	2	25	~4.5 × 10^1	Shin et al. (2009)
Sodium hydrosulfite	0.25	60	~1.4 × 10^3	Zhou et al. (2010)
Sulfur-containing compounds*	3	95	–	Chen et al. (2010)

*Sulfur-containing compounds include Na_2S, $NaHSO_3$, $SOCl_2$, SO_2, and $Na_2S_2O_3$.

Reduction of GO can be gained as well from photochemical reduction, thermal annealing at temperatures >1000°C, and electrochemical reduction (Cote et al., 2009; Zhou et al., 2009). One of the major drawbacks of GO is the very least electrical conductivity. With respect to the theory, GO turns into conducting when the functional groups approach 25%. After the elimination of the oxygen groups, *r*GO can be more graphitized from annealing at elevated temperatures. In this practice, those defects that remain after reduction are reorganized and the aromatic character of the monolayers grows. However, the existence of oxygen groups on the graphene surface is not ever unacceptable. Actually, the oxygen functional groups can be used for more functionalization of the layers by employing the well-established carbon chemistry for applications in gas sensors, catalysis, environmental remediation, and energy storage (Moon et al., 2010; Pei et al., 2010).

1.2.2. Isolation of Pristine Graphene Monolayers

The ultrasonication in organic solvents can achieve the exfoliation of graphite into single graphene layers. Unusual chemical conditions are provided by the acoustic cavitation because intensely high pressures and temperatures are reached in the liquid for short times (Suslick, 1990; An et al., 2010). If the solvent is capable to stabilize colloidal graphene because its surface energy is equal to that of graphene, and the free energy of mixing is non-positive, then the small graphite fragments intercalated by solvent molecules are formed and the graphitic basal structure is collapsed (Hernandez et al., 2008; Coleman, 2013). *N*-methyl pyrrolidone (NMP), dimethylformamide (DMF), *o*-dichlorobenzene, and pyridine, and other perfluorinated solvents were the initial solvents that are used successfully for this purpose (Figure 1.2).

A mixture of various derivatives is usually provided by the sonication methods where single graphene indicates a percentage of 1–15% and the rest involves few-layer graphene nanosheets, and the number of layers ranges from 2 to 10 (Blake et al., 2008; Bourlinos et al., 2009a). From selective centrifugation, the final percentage of single graphene layers can be raised. The benefit of these methods is that the graphitic character of the exfoliated layers is little affected as compared to in oxidation (Hamilton et al., 2009). However, violent collisions between particles are caused by the implosion of cavitation bubbles, at very high speed such that in air-saturated sonicated solutions, disengage the solvent and create peroxyl radicals (Misik and Riesz, 1996). Normally, the radical reactions are catastrophic and very useful in breaking C-C bonds (Guittonneauet al., 2010). As a result, prolonged sonication treatments lead to a higher number of defects mostly exists in oxidized carbon atoms at graphene edges in the form of carbonyl, carboxyl, and epoxy groups, and a reduction of the sheet size (Khan et al., 2010). With the addition of *N*-2-mercapto-propionyl glycine (tiopronin), a molecule that restricts reactions encourages by peroxides, radicals, and oxygen, such catastrophe while exfoliation of graphite in DMF can be markedly reduced (Dryer et al., 2011; Quintana et al., 2012). Another route to sonication is the procedure developed by Leon et al. (2011) which also has the upper hand of mitigating the progress of defects. This method is used for mechanochemical activation using ball-milling to peel graphite over interactions with melamine (2,4,6-triamine-1,3,5-triazine) under sturdy conditions (Moon and Gaskill, 2011).

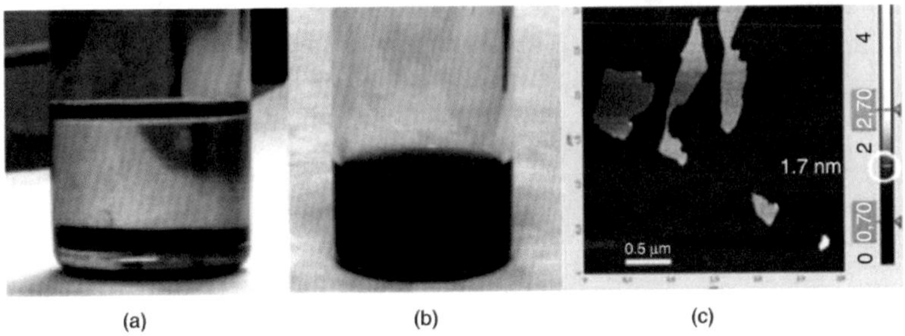

Figure 1.2. (a) In benzene, graphite precipitation after sonication; (b) in pyridine, partial exfoliation of graphite by sonication allows a dark colloidal dispersion for concentration 0.3 mg ml^{-1}; (c) AFM image of various pyridine-etched single graphenes layers.

Source: https://pubmed.ncbi.nlm.nih.gov/19408256/.

1.2.3. Large Scale Production of GO by Langmuir-Blodgett (LB) Methods

The Langmuir-Blodgett technique (LB) is used for the simple rules to deposit large GO flakes (5–20 μm) where a well dispersed and highly diluted water solution of GO is engaged as a subphase for the LB deposition (Cote et al., 2009).

A modification can be applied to the packing of the GO layers floating at the air-water interface by employing external pressure using the movable barriers of an LB trench. Opposite to hard and molecular colloidal particle monolayers, the GO single-layer flakes have a tendency to fold to oppose breaking down into multilayers. Cote et al. (2009) presented the first report of large-flake production of GO with the help of controllable deposition and the LB method. As shown in Figure 1.3 by controlling the surface pressure, high coverage of the GO sheets can be obtained, and the method is appropriate for large-scale production. By the reason of the injection of a long-chain molecule (e.g., octadecylamine) at the air-water interface, the GO sheets bind covalently and contribute to the formation of surfactant-GO layers (Gengler et al., 2010; Batzill, 2012). This hybrid Langmuir film can be moved to arbitrary support by horizontally decreasing the required substrate to associate the surfactant GO-water interface (higher hydrophobicity of the substrate gives rise to the quality of the deposited layer and the transfer ratio). This approach of transferring is called the Langmuir-Schaefer method (Sutter et al., 2008; Emtsev et al., 2009).

1.2.4. Other Methods of Graphene Production

Other methods for the production of single graphene layers from physicochemical and physical routes are less related to the current monograph and are thus only shortly cited in this section for completeness. Chemical vapor deposition (CVD) and thermal annealing of silicon carbide (SiC) can produce single graphene layers with promising results (Coraux et al., 2008; Kim et al., 2009). In graphene production, even various transition metals have been used as catalysts, but copper and nickel are the most promising, considering their low cost (Reina et al., 2009; Lee et al., 2014).

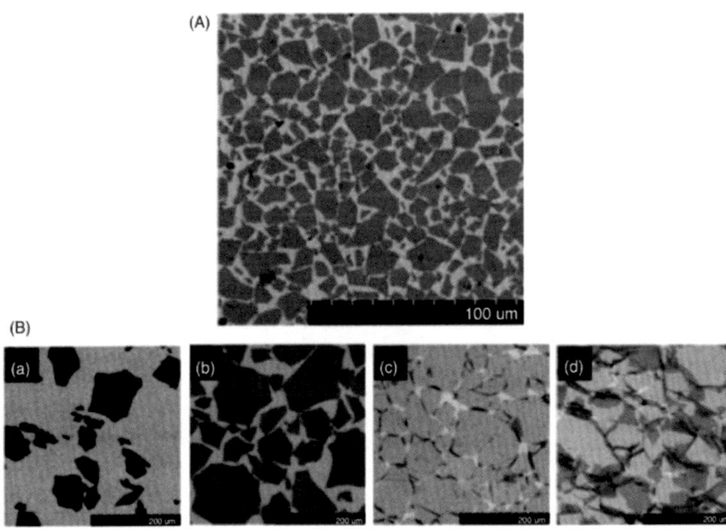

Figure 1.3. (A) SEM (scanning electron microscope) images of greatly covered graphene oxide monolayers, scale bar of 100 μm. Langmuir-Blodgett assembly of graphene oxide layers. (B) (a–d) SEM images of graphene oxide layers on a silicon wafer for various surface pressures. The packing density continuously raised by regulating the water interface pressure: (a) dilute monolayer of isolated flat sheets; (b) monolayer of close-packed GO; (c) overpacked monolayer with sheets folded at interconnected edges; and (d) overpacked monolayer with partially and folded overlap sheets.

Source: https://pubs.acs.org/doi/10.1021/ja806262m.

Whereas the thermal annealing of SiC at high temperatures range between 1000 and 1600°C head to the graphitization of the residue carbon atoms and the sublimation of silicon atoms. Another appealing scheme for the formation of graphene sheets with the predetermined size is the chemical unzipping of multiwalled carbon nanotubes (CNTs) which is known as *graphene nanoribbons* (GNRs) (Terrones et al., 2002; Mattevi et al., 2011). More accurately, strong oxidation or plasma etching cut the CNTs along their axis. Figures 1.4 and 1.5 show a method of presenting various procedures for the formation of GNRs also an atomic force microscopy (AFM) image of these graphene structures. The width of the GNRs is equal to the circumference of the nanotube, and their length is equal to the length of the nanotube. The edge structure (zigzag or armchair) broadly determines their electronic properties and, displays an energy gap for specific edge structures which gives rise to the falling width of the nanoribbon (Han et al., 2007).

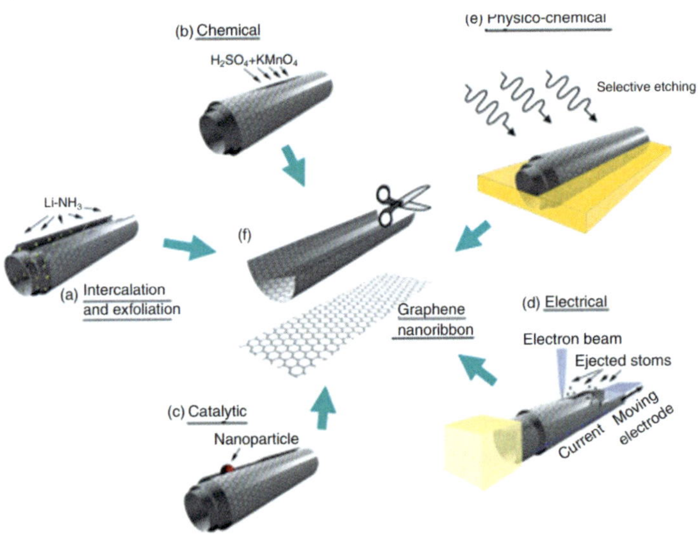

Figure 1.4. Diagrammatic representation of the different ways to generate graphene nanoribbons and unzip carbon nanotubes (Terrones et al., 2010).

Source: https://pennstate.pure.elsevier.com/en/publications/graphene-and-graphite-nanoribbons-morphology-properties-synthesis.

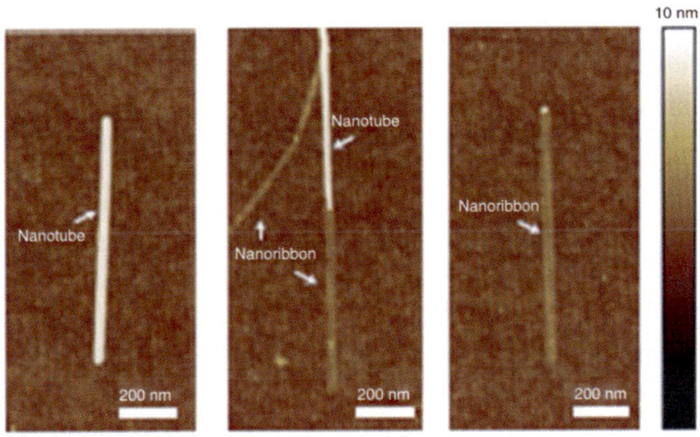

Figure 1.5. Feature AFM images of graphene nanoribbons by unzipping carbon nanotubes (Jiao et al., 2010).

Source: https://pubmed.ncbi.nlm.nih.gov/20364133/.

1.3.FUNCTIONALIZATION

Various reviews have already summarized the applications and synthesis of functionalized graphene materials (Gómez-Navarro et al., 2007; Tombros et al., 2007). However, some of them systematically reviewed the chemistry relevant to the functionalization of graphene. In this section, our focus will be on the structural and property changes of graphene upon functionalization, and the reaction mechanisms of graphene functionalization (Diba et al., 2016). The chemical functionalization of graphene through a covalent or noncovalent means is schematically illustrated in Figure 1.6 (Novoselov et al., 2005). The edge sites of a graphene sheet with dangling bonds are more reactive as compared to its basal plane. The dangling bonds can be utilized to covalently bond with several chemical moieties (see Figure 1.6(a)). These chemical moieties provide reactive groups for more modification or can increase the processability and solubility of graphene. The distortion of a p-p conjugation system is caused by the covalent functionalization of the graphene basal plane (refers to Figure 1.6(b)). On the other hand, the noncovalent functionalization keeps the electronic and atomic structures of graphene (see Figure 1.6(c)). The graphene can be imparted with particular supramolecular behavior (see Figure 1.6(e)) by the asymmetric functionalization on the two surfaces of a graphene sheet (see Figure 1.6(d)).

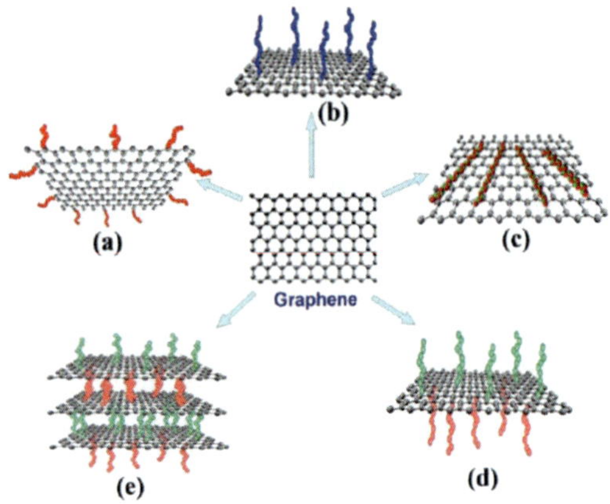

Figure 1.6. Chemical functionalization of graphene: (a) edge-functionalization; (b) basal-plane-functionalization; (c) noncovalent adsorption on the basal plane; (d) asymmetrical functionalization of the basal plane; and (e) self-assembly of functionalized graphene sheets.

Source: Source: https://pubs.acs.org/doi/10.1021/ar300122m.

1.3.1. Covalent Functionalization of Graphene Basal Plane

The basal plane of graphene is made up of sp^2 carbons, that is chemically unsaturated. Basically, there may be a chance to undergo covalent addition to transform the carbon atoms from sp^2 to sp^3 hybridization. In this procedure, the planar aromatic carbons change into a tetrahedral geometry with longer bonds. Geometric distortion is created by this functionalization, hence turns up high energy barriers. High energy reactants, e.g., fluorine atoms, hydrogen atoms, radicals, and strong acids are required by the covalent functionalization (Xie et al., 2015).

1.3.1.1.Hydrogenation

The most detailed studied covalent addition reaction on the basal plane of graphene is hydrogenation (Huang et al., 2011). Hydrogenated graphene has been prepared with the help of atomic hydrogen beams, where molecular hydrogen is broken on a hot filament, or through disclosure to hydrogen-based plasmas. This reaction transformed the hybridization of carbon atoms from sp^2 to sp^3, leading to the elongation of C-C bonds in the hydrogenated graphene. Hydrogen atoms turn to react with both sides of the basal plane of pristine graphene. If a single side is hydrogenated only, then the graphene sheet would revolve into a tube due to the unbalanced external stress (Abadal et al., 2017). Fully hydrogenated graphene is known as "graphane," and each carbon atom of graphane is covalently bonded to a hydrogen atom. As a result, the graphene layer was buckled. Figure 1.7 shows the schematic structures of three stable graphane isomers with a chair, stirrup, and boat configurations, together with two other isomers. It is indicated by the computational studies that the chair configuration is the most stable one. The H atoms are alternately adsorbed above and below the graphene sheet in this configuration (refers to Figure 1.7). The stirrup configuration is more stable than the boat configuration, where the stirrup configuration is also known as a washboard or zigzag configuration. The stirrup isomer incorporates varying zigzag chains with H atoms pointing down and up (see Figure 1.7).

The addition of hydrogen atoms has been expected to radically change the properties and electronic structure of graphene. Fully hydrogenated graphene is predicted to have an expanded bandgap. Different values were provided by various calculation methods (Morales-Torres et al., 2012).

Normally, a bandgap of 5.4 eV has been given by an adequate screened Coulomb interaction W (GW approximation) and the approximation of the Green function. Decreasing the degree of hydrogenation of graphene limits the bandgap. There are different optical properties of hydrogenated graphene from pristine graphene. As an instance, an absorption onset in the ultraviolet region is shown by the optical absorption spectrum of graphene, which makes it all transparent in the visible region. Magnetic properties are shown by the partially hydrogenated graphene (Sounas and Caloz, 2012).

1.3.1.2. Fluorination

Fluorination of graphene is like hydrogenation. A fluorine atom is linked to a carbon with a single bond; however, this bond has a stronger binding strength and a reversed dipole as compared with those of the C-H bonds in graphane. The binding energy of the C-F bond is less than that of the C-H bond; so, it would be simple to make a saturated fluoro-graphene as compared to the composition of a graphane. As a single side of graphene was disclosed to fluorine only, the fluorination was predicted to have maximum coverage of 25% (C_4F). An alternating conformation analogous to the chair conformation of graphene is the most stable structure for the double-sided fluorinated graphene (Meng et al., 2012). Fluorination of graphene can be achieved chiefly by two methods: (i) mechanical or chemical exfoliation of graphite fluoride, and (ii) treating graphene with an appropriate fluorinating regent (e.g., XeF_2). It is predicted by the theoretical calculations that fluorographene is electrically insulating with a minimum direct band gap of 3.1 eV (Novoselov et al., 2012). When electron-electron interactions are considered within the GW approximation, the calculated bandgap also increased to 7.4 eV. Same as to graphane, unique optical properties are possessed by fluorinated graphene. In contrast with graphene, higher transparency is shown for partially fluorinated graphene. Fluorographene shows up to be transparent at visible frequencies and only begins to absorb light in the blue region (Allen et al., 2010).

1.3.1.3. Oxidation

One of the most significant chemical reactions of graphene is oxidation. A more complex reaction is the addition of oxygen atoms to graphene rather than hydrogenation or fluorination because two covalent bonds can be formed by an oxygen atom instead of one. For the oxidation of graphene, there are three chemical routes that have been developed. The first is direct oxidation of graphene having strong oxidants, i.e., concentrated nitric acid, concentrated sulfuric acid, or potassium permanganate (Ferrari et al., 2006; Jinschek et al., 2011). The second is the oxidation of graphite, electrochemical methods, followed by exfoliation. Graphite flakes were collapsed into smaller fragments while the oxidation processes. The third oxidation process is unraveling CNTs and lengthwise cutting (Tuinstra and Koenig, 1970; Park et al., 2009).

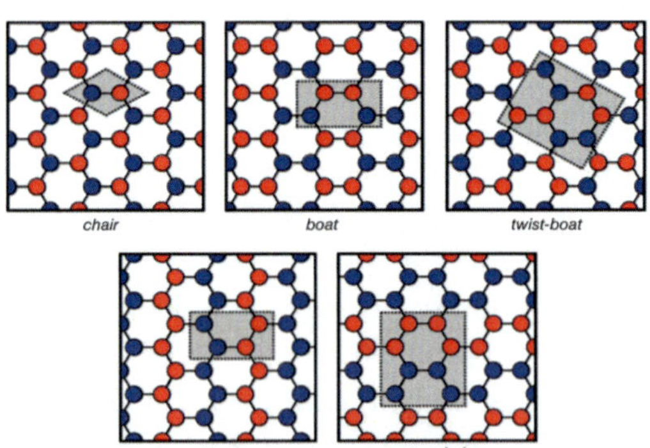

Figure 1.7. Five isomers of graphene where each carbon atom is equivalent. Red and blue colors represent hydrogen adsorption below and above the graphene layer.

Source: https://onlinelibrary.wiley.com/doi/abs/10.1002/wcms.1216.

One of the most popular graphene derivatives is GO which is an oxidation product of graphene. Basically, it is made up of about 45 mass% carbon. Even though various structure models have been presented, GO is a quite poly-dispersed material; it is very difficult to precisely define its exact structure. Lerf and Klinowski have proposed the broadly accepted structure model for GO (Mohiuddin et al., 2009). In the same way, the main functional groups on the basal plane of GO are hydroxyl and epoxy, and the carboxyl groups are chiefly situated at the edge. The solid-state nuclear magnetic resonance (NMR) spectra of GO have supported this structure (Lotya et al., 2010). It is indicated by the theoretical calculations that there is a 43 eV bandgap of GO with a saturated coverage of epoxy groups on its basal plane (Wang et al., 2008). The semi-metallic electronic structure of monolayer graphene can be revived if the top layer of bilayer graphene was decorated with epoxy groups. GO can be easily dispelled in aqueous media due to the appearance of oxygenated groups. GO can also be more functionalized using chemical reactions at its oxygen-containing groups (Bourlinos et al., 2009b).

1.3.1.4. Free Radical Addition

The inert chemical nature of pristine graphene can be repeatedly overcome by the free radical species. For free radical additions, the most commonly used adducts are aryl diazonium salts. Due to the p-electrons of graphene, it has an electron-rich basal plane. On the attack of an aryl radical on the basal plane of graphene, electrons can move from graphene to the radical (see Figure 1.9). Diazonium functionalization can open the zero-band gap of graphene. As an example, the reaction between (p-nitrophenyl) diazonium tetrafluoroborate and epitaxial graphene is reported by Rana et al. (2014), which results in functionalized graphene showing a bandgap of 0.36 eV. There is a conversion from sp^2 carbon to sp^3 by the covalent attachment of the aryl group on the basal plane of graphene. The conjugation length of the delocalized carbon lattice was modified as well by the free radical addition. Diazonium functionalization has successfully modified both pristine graphene and CMG at room temperature. After functionalization, there was a growth of the solubility of graphene in polar aprotic solvents (e.g., dimethylacetamide (DMAc), NMP, and DMF). Additionally, for more modification of graphene, there can be a reduction of the nitro groups of diazotized graphene into amine groups through the reaction of amine groups with carboxyl, acyl chloride, or hydroxyl groups of alternate functional components (Wei et al., 2013).

1.3.1.5. Cycloaddition Reaction

Cycloaddition reactions are different from other typical organic reactions because cations or anions are not produced by these reactions as intermediates. Alternatively, the electrons move in a circular manner that includes concurrent bond formation and bond cleavage. In accordance with the atom number of the addition ring, there are four types of the cycloaddition functionalization of graphene: (i) [2+1] cycloaddition forming a three-membered ring, (ii) [2+2] cycloaddition forming a four-membered ring, (iii) [3+2] cycloaddition forming a five-membered ring, and (iv) [4+2] cycloaddition forming a six-membered ring (Figure 1.8) (Zhou et al., 2014).

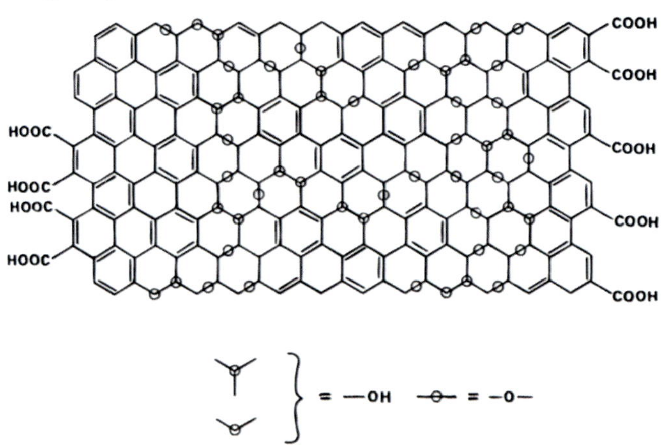

Figure 1.8. Lerf-Klinowski structure model of GO.

Source: https://pubs.acs.org/doi/10.1021/jp9731821.

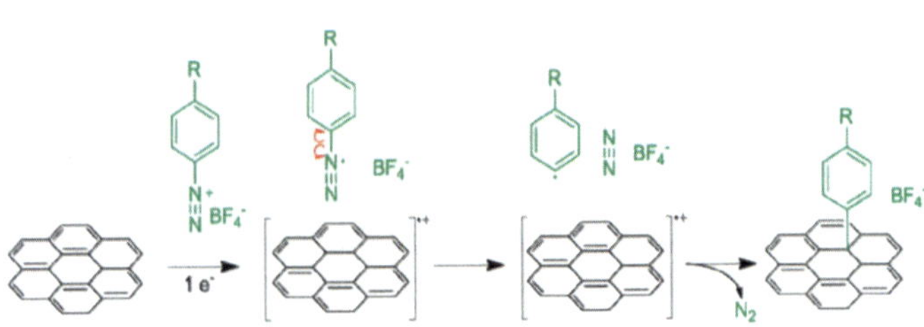

Figure 1.9. The technique of the free radical addition for derivatives of phenyl on graphene.

Source: https://pubs.rsc.org/en/content/articlelanding/2013/cs/c2cs35474h.

One of the earliest used methods to functionalize the sp^2 carbon network of graphene is the [2+1] cycloaddition reaction. Nitrene and carbene are two classic intermediates for this purpose. For example, dichlorocarbene has been well coupled to the sp^2 carbon network of graphene (see Figure 1.10(a)) due to its high reactivity. In this case, there is a spontaneous reaction of the singlet carbene (an electrophile) with the sp^2 carbon atoms; there is an interaction of the electron pair of carbene (HOMO) with the p* antibonding orbital (LUMO) of CQC, and interaction of the empty p orbital of carbene (LUMO) with the p bond (HOMO) of CQC. Hence, the dichlorocarbene addition changes the electronic characteristics of graphene from metallic to semiconducting and disturbs the p conjugation of graphene. Moreover, the addition of polar chlorine atoms gives rise to the solubility of graphene in organic solvents. The energy gap of graphene can be tuned by the cyclopropane adduct. Same as to carbene, nitrene intermediates results in the composition of aziridine adducts on graphene (refers to Figure 1.10(b)).

In the cases of using benzyne or aryne as the reactive species, four-electron cycloadditions appeared on the graphene sp^2 carbon network from an elimination-addition operation (Wang and Shi, 2015).

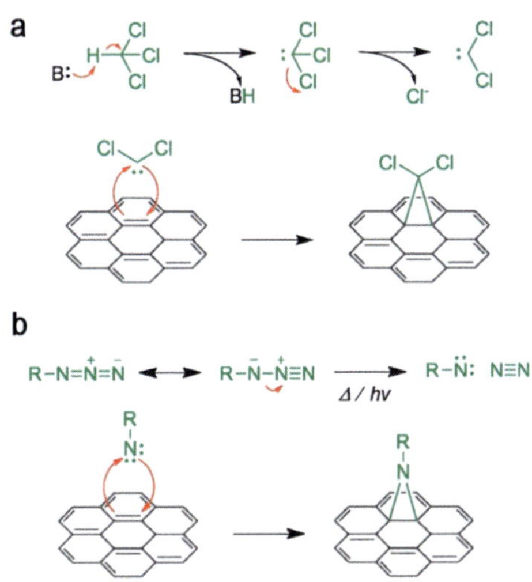

Figure 1.10. (a) The technique of the formation of dichlorocarbene with cyclopropanation of graphene with dichlorocarbene (bottom), and chloroform and base (top); (b) Method of the formation of nitrene through cycloaddition of nitrene onto graphene (bottom), and decomposition of azide (top).

Source: https://pubs.rsc.org/en/content/articlelanding/2013/cs/c2cs35474h.

As an example, fluoride-induced benzyne, the CQC bond is attacked by the electrophilic benzyne on the graphene basal plane, leading to a [2 + 2] cycloaddition (refers to Figure 1.11(a)). A five-membered ring can be accomplished froma six-electron cycloaddition between the sp^2 carbon atoms and a 1,3-dipole on graphene (see Figure 1.11(b)).

Moreover, a six-membered ring can be acquired via the Diels-Alder cycloaddition, which is a very popular pericyclic reaction in organic chemistry. The interaction between a dienophile and a conjugated diene is involved by this [4 + 2] cycloaddition reaction; the LUMO of dienophile and the overlap of the HUMO of diene results in the formation of a six-membered ring (see Figure 1.11(c)). The Diels-Alder cycloaddition is facilitated by the unsaturated dangling bonds located on C atoms.

1.3.2. Asymmetrical Functionalization of Graphene Basal Plane

Graphene is made up of carbon atoms that are linked together by strong covalent bonds, which are expected to be impermeable to molecules or small atoms. This characteristic of graphene makes it potential for bifacially asymmetric modification. As an example, Janus graphene has been synthesized successfully, where Janus graphene consists of two types of functional groups detached by the single-mediated carbon layer (expressed as X-G-Y, refers to Figure 1.12). In this case, there is a utilization of a 200–300 nm thick poly(methyl methacrylate) (PMMA) film for a two-step functionalization as a flexible macroscopic mediator. The asymmetrically bonding X and Y atoms on both surfaces of graphene have obtained four types of Janus graphene on the combination of fluorination, photochlorination, diazotization, phenylation, and oxygenation reactions. The functionalities are able to influence the surface wettability on one side and the chemical reactivity on the opposite side. So, the self-assembly of graphene sheets can be efficiently controlled by this asymmetric structure (Tian et al., 2014).

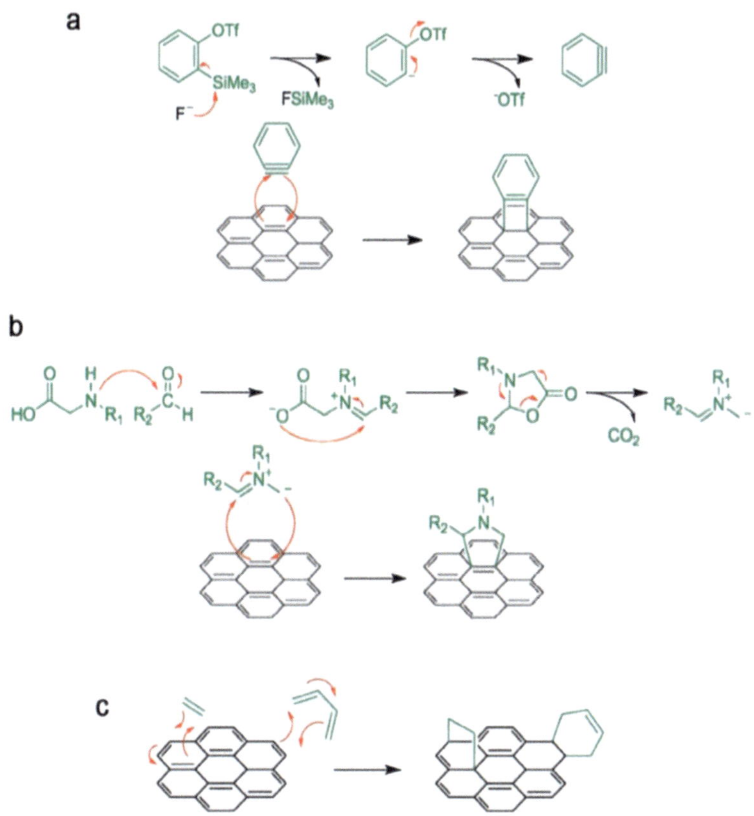

Figure 1.11. (a) Method of the formation of benzyne by means of the cycloaddition of graphene with benzyne (bottom), and a fluoride-induced decomposition methodology (top); (b) the technique of the formation of azomethine ylide via the 1,3-dipolar cycloaddition of graphene with azomethine ylide (bottom), and N-methyl glycine (top); (c) the strategy of Diels-Alder cycloaddition with graphene as a dienophile and diene.

Source: https://pubs.rsc.org/en/content/articlelanding/2013/cs/c2cs35474h.

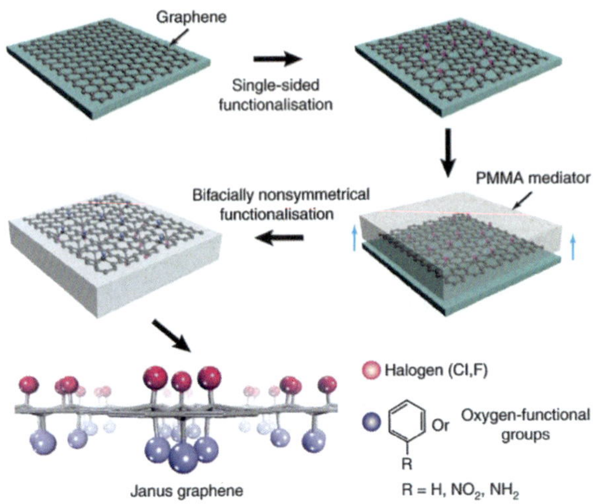

Figure 1.12. Schematic demonstration of the PMMA-mediated transfer procedure to fabricate Janus graphene.

Source: https://www.nature.com/articles/ncomms2464.

1.3.2.1. Functionalization of Graphene Edges

The tetrahedral geometries are adopted by the edge carbons of graphene that provides them more freedom as compared to that of basal plane carbons without leading extra strain. Hence, as compared to the edge carbon atoms on the basal plane, those are more reactive here. There are two typical configurations of graphene edges: (1) "armchair," and (2) "zigzag."

They can also have a consolidation of both configurations. There is an unpaired electron of each carbon atom of the zigzag edge which makes it clear to bond with other moieties. As the existence of a triple covalent bond within the open edged carbon atoms, there are more stable carbon atoms of the armchair edge. Moreover, the chemical reactions are facilitated by the existence of edge defects in CMG, and offers convenience for functionalization. The typical defect sites for functionalization consist of vacancies, edge sites with dangling bonds, and defect sites decorated with functional groups, i.e., epoxy, carboxyl, and carbonyl groups. The assembly and the solubility behavior of graphene can be improved by the functionalization of graphene edges. Still, no obvious changes are exhibited by the bandgap of graphene, as its sp^2 network is undisturbed (Wu et al., 2013).

1.3.2.2. Functionalization of Pristine Graphene Edge Sites

There can be a quick reaction of the termination-free graphene edges with adsorbed molecules under ambient conditions. The free zigzag sites are the carbene type, whereas the free armchair sites are the o-benzyne (or carbyne) type in terms of modern organic chemistry (refers to Figure 1.13). Many carbene-related reactions like insertion and cycloaddition, are enabled by the zigzag sites. The armchair-edged GNRs can also apply insertion and pericyclic reactions. Currently, Diels-Alder reactions have been achieved at graphene edges, and it is becoming a strong approach to tune the electronic characteristics of graphene under mild conditions. The section "Functionalization on graphene basal plane" have already discussed the technique of these addition reactions (Kong and Chen, 2014).

1.3.2.3. Functionalization of CMG Edge Sites

Several moieties like carboxylic, hydroxyl, and carbonyl groups can decorate the edge sites of CMG. These groups improve the solubility of CMG in aqueous or organic solutions and aid CMG for further functionalization.

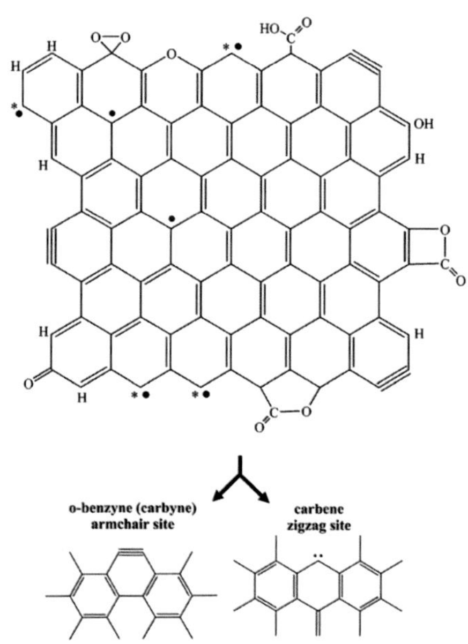

Figure 1.13. Schematic illustration of the main chemical features in a graphene sheet, along with its typical surface functionalities, containing the free edge sites.

Source: https://pubs.acs.org/doi/10.1021/ja050124h.

GO is the most generally used functionalization precursor with abundant carboxyl groups at its edges (see Figure 1.8). Typically, the carboxyl groups of GO have to be triggered via two routes: (i) coupling with amine or alcohol and (ii) pre-treatment with $SOCl_2$. These methodologies are not restricted to GO; they can also be utilized to activate oligomers, small molecules, C_{60}, and polymers.

For the modification of CMGs, the amidation reaction of coupling amino and carboxyl groups is a powerful route. Using this approach chromophores, amino-terminated polymers, ligands, and biomolecules have been anchored successfully upon CMG sheets. Likewise, the coupling of carboxyl and alcoholic groups has been broadly used to functionalize CMGs as well. A significant reaction is within GO and hydroxyl-terminated poly(3-hexylthiophene) (P3HT). The P3HT-grafted GO is soluble in trivial organic solvents, easing device fabrication, and materials analysis through solution processing. As an example, with the help of P3HT-grafted GO along with C_{60}, a photovoltaic device can be designed. The power conversion efficiency (PCE) was shown to be improved for this device as the extended electron delocalization of GO onto covalently associating P3HT (Lu et al., 2012).

1.3.3. Noncovalent Functionalization of Graphene

Noncovalent functionalization normally covers decorating functional species onto graphene sheets by the way of hydrophobic, p-p stacking, hydrogen bonding, or/and electrostatic interactions. This is a physical process accomplished by adsorption of surfactants, polymer wrapping, or small aromatic molecules, etc. (see Figure 1.14). The structure and characteristics of graphene are maintained noncovalent functionalization of graphene opposite to the covalent functionalization (Kim et al., 2013).

1.3.3.1. Conjugated Compounds

Irregular aggregates in solutions are easily formed by the rGO and pristine graphene due to their hydrophobicity and conjugated structures. To acquire stable suspensions of graphene sheets, decorating graphene with conjugated compounds can be employed. Normally, the conjugated compounds have functional groups and polyaromatic rings. The functional moieties stabilize graphene sheets in solvents or/and provide graphene with new functionalities, and the conjugated polyaromatic rings connect to the sp^2 network of graphene sheets through p-p stacking. Classical conjugated compounds are anthracene, naphthalene, porphyrin, pyrene, and their derivatives, etc. As an example, upon rGO sheets in a monolayer from electrostatic interactions and p-p stacking, 5, 10, 15, 20-tetrakis (1-methyl-4-pyridinio) porphyrin (TMPyP) molecules can be decorated. The firm interaction between these two elements forces the molecular flattening of TMPyP molecules. The TMPyP functionalized rGO sheets can be stably break up in the water and can be utilized to detect Cd^{2+} ions selectively and rapidly.

1.3.3.2. Polymers

From noncovalent functionalization with polymers, stable dispersions of graphene sheets in several solvents can be obtained as well. For example, there can be electrostatically interaction between the amine-terminated polystyrenewith the carboxylate groups of rGO sheets. This interaction allowed the coating of polystyrene chains onto rGO sheets, changing rGO sheets from an aqueous phase to an organic phase. There is also an attraction of a great deal of interest in immobilizing biomacromolecules on graphene sheets from noncovalent functionalization. The applications of biomacromolecules are extended by the excellent properties, large specific surface area, and biocompatibility of graphene (Wu et al., 2015). As an example, rGO sheets were adorned with amphiphilic polyethylene glycosylated (PEGylated) polymers, translating them stable in biological systems. Alternate conventional biomolecules that can immobilize onto graphene sheets involve chymotrypsin, enzymes, trypsin, proteins, etc.

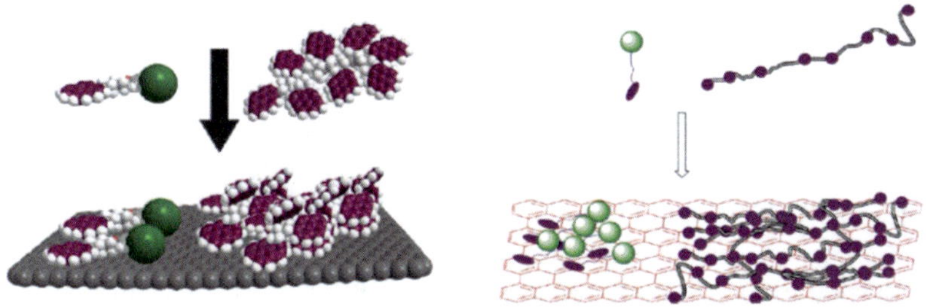

Figure 1.14. Demonstration of the noncovalent functionalization of graphene using small molecules and polymers.

Source: https://pubs.acs.org/doi/full/10.1021/jz4010448?src=recsys.

1.4. DOPING

The most promising technique to adapt the electronic structure of graphene is chemical doping with the help of charge extraction or injection. Pristine graphene refers to a zero-band gap semimetal, and its Fermi level is situated near the Dirac point (refers to Figure 1.15(A)). For the tuning of the bandgap, chemical doping is introduced by shifting the Dirac point corresponding to the Fermi level. The doped graphene is a p-type (n-type) semiconductor because the Dirac point is above (below) the Fermi level (see Figure 1.15(B)). By standard, there are two classes of chemical doping of graphene: (i) surface transfer doping, and (ii) substitutional doping (Kosidlo et al., 2013).

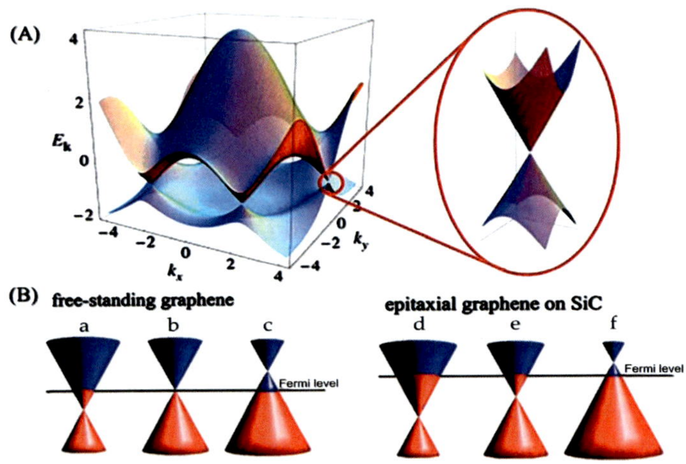

Figure 1.15. (A) Right: zoom-in of the energy bands close to one of the Dirac points. Left: electronic dispersion in the honeycomb lattice. (B) A schematic representation of the location of the Fermi level as a function of doping and the Dirac point. Right: (d) n-type doped, (e) pristine, and (f) p-type doped epitaxial graphene grown on silicon carbide. Left: (a) n-type doped, (b) pristine and (c) p-type doped freestanding graphene.

Source: https://pubs.rsc.org/en/content/articlelanding/2011/jm/c0jm02922j.

Surface transfer doping is carried out by the charge transfer (CT) within the dopants adsorbed on its surface and the graphene. In most cases, the chemical bonds of graphene are not destroyed by this route. As for substitutional doping, the carbon atoms present in the skeleton of graphene are replaced by the heteroatoms (e.g., boron, sulfur atoms, and nitrogen), disturbing its structure (Li et al., 2011).

1.4.1. Surface Transfer Doping

In surface transfer doping, molecules with donating groups or electron-withdrawing are adsorbed on the surfaces of graphene sheets, resulting in the formation of n-type or p-type doped graphene. A collection of species can adsorb onto graphene sheets, e.g., organic molecules, gases, and metal atoms. In an oxygen or air atmosphere, the p-type doped graphene can be easily composed. It is reported by Kotal et al. (2016) that with the adsorption of water on the surface of the graphene sheet, the bandgap of graphene can be opened. Strong electron acceptors like Br_2, I_2, and NO_2, can also act as p-type dopants. Whereas NH_3, CO, and ethanol are n-type dopants (Lu et al., 2013).

The tuning of the bandgap of graphene also makes the adsorption of organic molecules able. For example, the graphene is doped by Wei et al. (2013) with 3-dimethyl-2, 2-(2-methoxyphenyl)-1, 3-dihydro-1H-benzimidazole (o-MeO-DMBI). With the change of the amount of o-MeO-DMBI, the electrical property of graphene has been transformed from p-type to n-type (see Figure 1.16).

Various metal atoms can be absorbed by the surface of graphene. Because of the difference between the work functions of these two components, the doping types of graphene can be adjusted. At the time of the transfer of an electron to equilibrate the Fermi levels, graphene is p-doped with Pt and Au, while n-type doped with Ag, Cu, and Al (Kim et al., 2014).

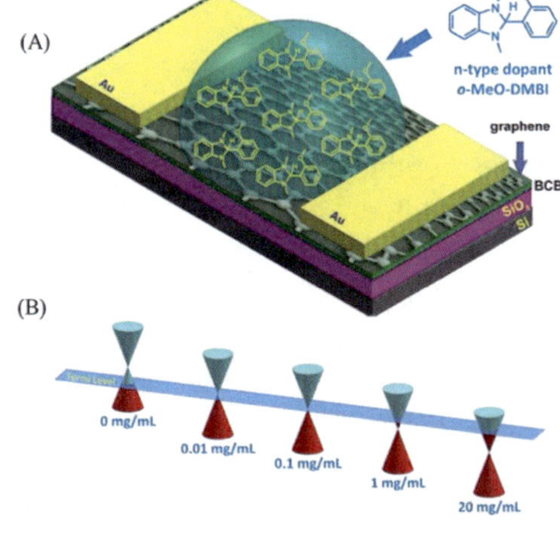

Figure 1.16. (A) Chemical structure of o-MeO-DMBI and the schematic demonstration of a o-MeO-DMBI doped CVD-grown graphene transistor by the so-

lution process. (B) Schematic representation of the shifts in the fermi level approaching the Dirac point varied with the o-MeO-DMBI solution concentration.

Source: https://pubmed.ncbi.nlm.nih.gov/23537351/.

1.4.2. Substitutional Doping

A diverse range of atoms have been introduced into graphene to broaden its applications and transform its electron density. Atoms with hardly one or more valence electrons as compared to that of carbon results in the formation of n-type or p-doped graphene. For the preparation of heteroatom doped graphene, two chemical approaches have been produced: in post-treatment and situ doping. The effective techniques for in situ doping of graphene at the time of its growth are solvothermal reaction, arc discharge, and CVD. Homogeneously doped graphene materials can be produced using these techniques. Plasma in atmospheres with molecules incorporating heteroatoms or post-treatment of bulk graphene materials using thermal annealing likely dopes them only on their surfaces. The charge distribution of carbon atoms in graphene will be influenced by the heteroatom doping which makes the doped graphene a good electrode material in consideration of the fabricating field-effect transistor (FET) devices. Furthermore, the structural defects happened by doping in graphene are normally the active sites for surface chemical reactions. Hence, the superior performances are exhibited in sensors and catalysis by the doped graphene materials frequently (Liu et al., 2013).

1.4.2.1. Nitrogen-Doping

There are mainly three bonding configurations of the nitrogen atoms (N) doped in the carbon lattice of graphene: pyridinic N, pyrrolic N, and quaternary N (or graphitic N) (see Figure 1.17). Particularly, two C atoms at the defects or edges of graphene are bonded to the pyridinic N atom. The pyridinic N atom bonds one oxygen atom in pyridinic N oxide. A pyrrolic N regards the formation of a five-membered ring and N atom bonding with two C atoms, the same as to that in pyrrole. Quaternary N corresponds to the three N atoms that uniformly place C atoms in a hexagonal ring. Within these bonding configurations, pyrrolic N is sp^3 hybridized, and quaternary N and pyridinic N are sp^2 hybridized. There is a marginal influence of pyridinic and quaternary N on the graphene structure. Comparing, the planar structure of graphene is disrupted by the sp^3 hybridized pyrrolic N.

The types of defects are indicated by the first-principles calculations on the reaction paths of NH_3 on graphene defects that figure out the configurations of doped N atoms: pyridinic or pyrrolic-N at divacancies, graphitic-N at single vacancies, N in a four-membered ring at zigzag edges, and pyrrolic-N at armchair edges. Literally, with the introduction of defects into graphene sheets through a physical approach (e.g., N^+-irradiation) and taken place by annealing in NH_3, N-doping can be controlled (Mukai et al., 2009).

The electro-negativity of N (i.e., 3.04 on the Pauling scale) is greater than that of C (i.e., 2.55 on the Pauling scale). Therefore, the polarization in the carbon network is created by the N-doping, with that influencing the magnetic, optical, and electronic properties of graphene. A bandgap is opened by the N-doping near the Dirac point, exploring graphene with semiconducting properties. The semiconducting property of N-doped graphene is dependent on the configurations of doping atoms. For a graphitic N (refers to Figure 1.17), one electron is hooked in p-bond formation, the 5th electron has participated in the p* state, and three valence electrons of N are bonded to the adjacent carbon atoms. Hence, electrons are contributed to N atoms to the graphene lattice, leads to an n-doping effect. In contrast, with the withdrawal of electrons from the graphene sheet, the p-doping effects are induced by the pyrrolic and pyridinic N atoms (Kim et al., 2014).

The work function of graphene can be effectively tuned by N-doping that is helpful for (light-emitting diode) LEDs and FETs. The work function of graphene doped with pyridinic N (4.83 eV), hydrogenated pyridinic N (4.29 eV), and graphitic N (3.98 eV), and pristine graphene (4.43 eV) was by different scientists (Li et al., 2011; Lu et al., 2016). The electron-accepting or donating nature of every N-bonding configuration causes the change in graphene work function.

At room temperature, magnetic hysteresis is shown by the pristine graphene; however, a magnetic moment can be created by the heteroatom doping in graphene. The generation of a magnetic moment is not being able by the graphitic N (see Figure 1.17) among the doping N atoms, due to the deficiency of nonbonding electrons. For pyridinic N, the unpaired spins are situated on N atoms that focused on graphene edges. Therefore, it has less impact on the spin polarization of the edge states. Hence, strong magnetic moments can only be created by pyrrolic N atoms.

The optical properties of graphene sheets can also be tailored by N-doping. The control of N-doping on the photoluminescence (PL) feature of graphene was studied by Batzill (2012). From ground N 1s orbitals to the unoccupied 2p orbitals (p* state), the electrons were excited when illuminating N atoms with light. Then, from the p* state to the p state, the electrons were conveyed together by energy discharged in the form of PL emission. So, the PL emission of graphene is enhanced by N-doping essentially.

The CVD technique can prepare the large-area high-quality N-doped graphene. Commonly, to form N-doped graphene, a mixture of an N-containing gas (e.g., NH_3) and a carbon source gas (e.g., CH_4) break down at a metal catalyst (e.g., Cu or Ni) foil. Solid and liquid organic precursors (e.g., pyrazine, pyridine) have also been probed for this reason. Thermal annealing of GO in the existence of various N sources or solvothermal treatment can prepare N-doped graphene with high N content.

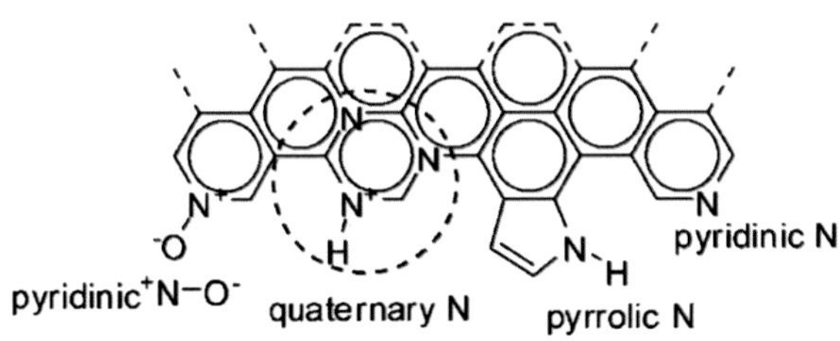

Figure 1.17. In N-doped graphene, bonding configurations for nitrogen atoms.

Source: https://link.springer.com/article/10.1007/s11244-009-9289-y.

1.4.2.2. Boron Doped Graphene

Boron ($2s^1 2p^1$) is the adjacent element of carbon ($2s^2 p^2$) having only one less valence electron; in this manner, it is a usual p-type dopant for graphene. In-plane substitutional doping (e.g., in-plane BC_3) is more stable than out-of-plane bonding (refers to Figure 1.18(a)). The planar structure of graphene is maintained because a B atom model sp^2 hybridization in the carbon lattice. As compared to this graphitic bonding configuration, structural distortion will be created by the bonding of a B atom in a vacancy. A novel type of structural rearrangement is suggested by the theoretical calculations, which is a tetrahedral-like BC_4 configuration from which all the dangling carbon atoms are saturated (refers to Figure 1.18(b)). The planar structure of graphene is distorted by this B-doped configuration. The downshift of the Fermi level of graphene is induced by the B-doping towards its Dirac point. Theoretical calculations represent that by doping a B atom into a graphene network with 50 carbon atoms, a bandgap of 0.14 eV can be proposed.

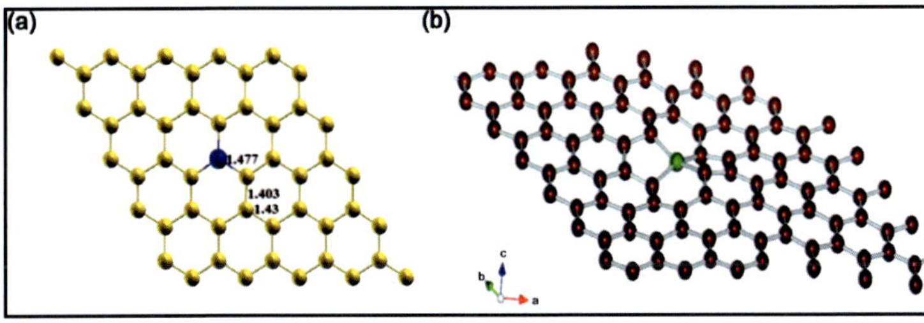

Figure 1.18. (a) Substitutional doping of B (blue ball); (b) B atom in a vacancy with the symmetric disposition (i.e., green ball).

Source: https://pubs.acs.org/doi/abs/10.1021/jp106764h.

With the use of boron powder and ethanol as the boron and carbon sources, the B-doped few-layer CVD graphene was formed. Approximately 0.5% B atoms are contained by the graphene, exhibiting the electrical properties of a p-type semiconductor. Moreover, graphene doped with 3.5 at% B was created through thermal annealing of GO in the existence of boron oxide. In contrast to pristine graphene, there are more excellent electrocatalytic characteristics of B-doped graphene. With the modulation of the doping level, the bandgap of B-doped graphene is adjustable. As an example, on changing the B content from 0 to 13.85% by a microwave plasma method, the bandgap of doped graphene transformed from 0 to 0.54 eV. The on-surface chemical reaction of an organoboron compound synthesized the Boron doped GNRs with widths of 7, 14, and 21 carbon atoms. In the centers of the nanoribbons, the locations of doped B atoms are explained with a programmed content of 4.8 atom%. An effective approach was provided by this work for doping graphene with B atoms in a controllable trend.

1.5. CHARACTERIZATION OF GRAPHENE

Opportunities are offered by the isolation of single graphene sheets for its examination by numerous microscopic and spectroscopic techniques; samples can be either in the form of graphene sheets deposited on the proper substrates or dispersion. The most generally used characterization tools are presented in this section. As for most AFM and nano-materials, electronic microscopies are effective tools for the characterization of graphene derivatives and graphene. Single-layer graphene from a few-layer and double-layer graphene can be distinguished by the Spectro-microscopy and Raman spectroscopy, and the clear manifestation is given on the number of defects exist in the material. To discern changes in the structure of graphitic materials after and before the functionalization of graphene sheets, thermogravimetric analysis (TGA) diagrams are beneficial. A single graphene layer can be visualized by the optical microscopy that is located on the right substrate. X-ray diffraction (XRD) update on the achievement of intercalation or exfoliation of graphite and is specifically helpful to illustrate functionalization.

1.5.1. Microscopic Observation

The spin coating or drop-casting an eminently diluted graphene dispersion on a silicon wafer often perform the characterization of a graphene material by AFM because these deposits are flat enough to permit for the recording of height differences on the monoatomic level. It is shown in Figure 1.19(A) shows representative AFM images of single graphene sheets. The height differences between the folded and flat part of the graphene layer can be clearly observed. When decorated with hydro- and oxygen-containing groups, the average height of annealed single-layer graphene flakes is commonly in the range of 0.8–1.2 nm. The average height of the flakes decreases to 0.3–0.5 nm after the graphitization treatments at high temperature, displaying the "fingerprint" of a single atomic sheet identical to the mechanically exfoliated flakes (Gómez-Navarro et al., 2007).

Figure 1.19(B)-(a) and (b) presents the conventional TEM (transmission electron microscope) images of single-layer graphene; the film is nearly transparent. It is shown in Figure 1.19(B)-(c) that on recordation from aberration-corrected instruments, the defect structures at grain boundaries can be viewed with atomic resolution. The exit wave reconstruction visualizes the atomic structure of graphene sheets which is a state-of-an-art TEM technique, where 10–30 HR-TEM images are captured at various defocus values and consolidated into the complicated wave of electrons at the exit plane of the sample. Figure 1.19(B)-(d) shows an example of a phase image of such an exit wave of electrons remains in a graphene sheet. Comparing to the single high-resolution TEM images, phase images permit to determine single and double graphene layers and allow for quantitative analysis of the contrast. The inset of Figure 1.19(B)-(d) shows a defect-free graphene lattice, in which the location of the individual carbon atoms can be known. A single graphene sheet is clearly indicated by this image, as the AB stacking of a double sheet would result in the existence of more atoms in the center of the hexagons. It is also indicated by the overview image shown in

Figure 1.19(B)-(d) that adsorbents are likely to be available at the surface of the graphene layer, increasing the ripple-like contrast present in this image.

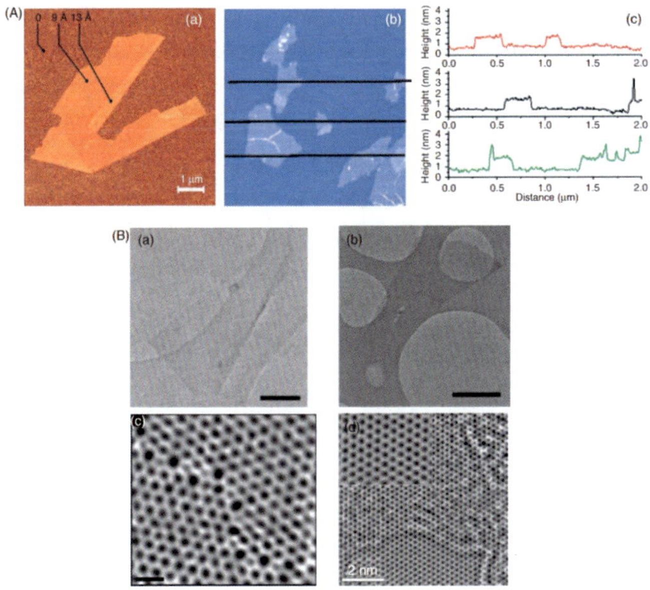

Figure 1.19. (A) (a) AFM image of the pristine single graphene sheet. The height which correlates to the thickness of a single layer is 0.9 nm whereas a folded sheet is measured at a height of 1.3 nm; (b) AFM image and height profile of a single GO layer; (c) height profile collected from the lines marked in black on the AFM micrograph. (B) TEM images of (a, b) pristine single graphene sheets; (c) aberration-corrected TEM image: grain boundary in a single graphene sheet; (d) HR-TEM of a graphene monolayer generated by exfoliation of graphite in the existence of tiopronin as radical trap.

Source: https://pubs.rsc.org/en/content/articlelanding/2012/cc/c2cc35298b.

1.5.2. Raman Spectroscopy

Raman spectroscopy is a broadly utilized tool for the characterization of carbon materials; it is especially useful for determining the structure of graphene nanosheets concerning the number of graphene layers, also the extent of functionalization, and the presence of defects. Figure 1.20 shows the pioneering study by Ferrari et al. (2006) of the Raman spectrum of pristine single graphene which described how for few-layer graphene including one to five layers, the actual number of layers can be taken from the spectrum.

There are two characteristic features of the Raman spectrum of a pristine single graphene layer, i.e., at 1580 cm^{-1} called the *G band* and at 2700 cm^{-1} known as the *G′band*. The consequence of the doubly degenerate zone center E_{2g} mode is the G band. The G band carries testimony as well to the number of layers. With the increase in the layer thickness, the band position moves to the lower wavenumbers agreeing to the determined positions for these band locations. The position of the G band is too sensitive to doping and strain results in the breaking of this band. The second order of zone boundary phonons is the G′ band and is usually called the *2D band.* When graphene has an adequate number of defect sites, the first order of the zone boundary phonons is noticed only as a peak about 1350 cm^{-1}, known as *the D band.* In the case of a pristine graphene monolayer generated by micromechanical cleavage, such a band is not identified due to the deficiency of defects.

As shown in Figure 1.20(c), the G′ peak varies with the number of layers: a sharp symmetrical peak below 2700 cm^{-1} is referred to the G′ peak of a single layer of graphene. This peak is moved to lightly higher wavenumbers for the bilayer graphene and turns to the vast with a shoulder toward lower wavenumbers. The peak shifts to higher wavenumbers with the increase in the number of the layers, and at last it appears in a five-layer nanosheet as a broad double peak where both components have a $1/2$ ratio (refers to Figure 1.20(c)). The G′ band is identical to that of a sample with five layers for a nanosheet with more than five layers.

For large pristine single graphene sheets that are processed by micro-mechanical cleaving without flaws, Raman spectra without D band are hardly noticed. In many cases, there is enough number of defects in the pristine graphene sheets to bring some D band intensity. There is a direct dependence on the height of the D band and the number of the sp^3 carbon atoms of the graphene surface, and therefore, it depends on the number of flaws of the graphene nanosheets. As concerns the quality of graphene, the D band is an expression for the "quality" of the graphene nanosheet and the aromatic character and corresponds to the initiating material and the production method. For example, it is reported in Figure 1.21(A) that spectra corresponding to the graphene sheets formed by the depilation of graphite in water and balanced with a surfactant which represent a powerful D band that remains even after strengthening at 500°C. An identical strong D band is recognized in the Raman spectra of GO sheets. Here the D band is a general property of the Raman spectra from the presence of sp^3 carbon atoms in the graphitic surface moves along with the production of oxygen groups (refers to Figure 1.21(B)).

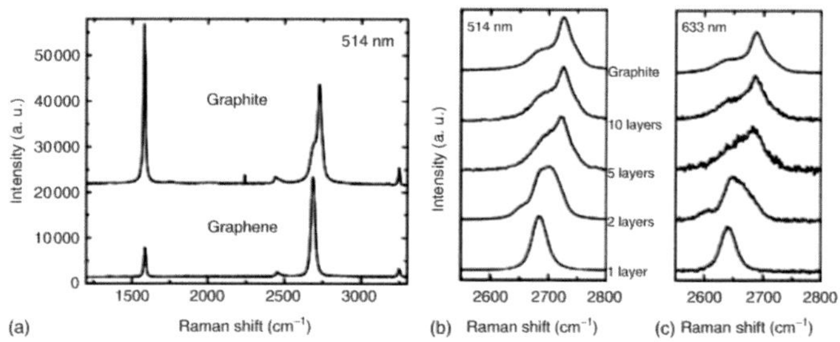

Figure 1.20. (a–c) The Raman spectra of pristine graphene in contrast to that of graphite and the G′ band of different multilayered graphene nanosheets.

Source: https://journals.aps.org/prl/abstract/10.1103/PhysRevLett.97.187401.

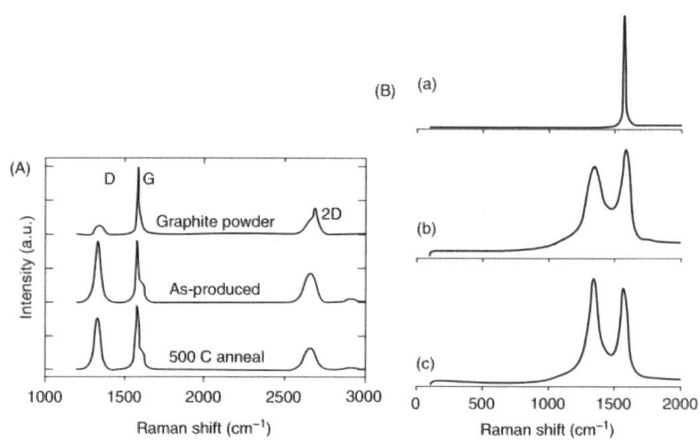

Figure 1.21. (A) Raman spectrum of graphene as produced and after annealing at 500°C in contrast to that of the starting graphite; (B) the Raman spectra of graphite (a); GO (b); and the reduced GO (c).

Source: https://www.scholars.northwestern.edu/en/publications/synthesis-of-graphene-based-nanosheets-via-chemical-reduction-of-.

1.5.3. Thermogravimetric Analysis (TGA)

Writings regarding the characterization of graphene nanosheets and its derivatives generally involve TGA because of the structural variations of graphitic materials after and before procedures, i.e., the functionalization of graphene sheets or the exfoliation and oxidation of graphite increase the apparent differences in the mass loss being a function of increasing temperature (with fixed heating rate). For example, the TGA data for GO, *r*GO, and graphite as presented by Wang et al. (2008) is reported in Figure 1.22. When the sample is heated in air, the explosion of graphite initiates at 650°C, whereas 20% of the weight is loosed by the GO at 200°C and is ultimately decomposed at 550°C. The elimination of the oxygen groups is connected to the first weight loss of GO, whereas the lower thermal stability of GO is demonstrated by the lower combustion temperature of GO as compared to graphite due to the existence of defects that appeared after removal of oxygen functional moieties. In *r*GO, an intermediate thermal behavior is observed which reflects the lower number of oxygen groups in the present material. For both *r*GO and GO, the exfoliation also influences the lower combustion temperature, which makes the sheets more feasibly available to air as compared to when they are firmly packed in graphite.

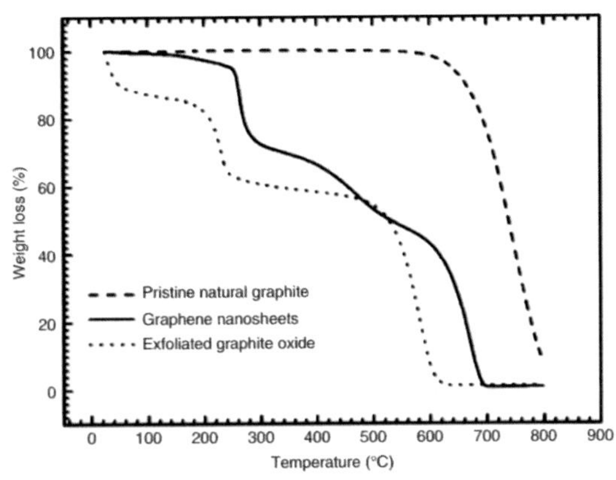

Figure 1.22. TGA curves of natural graphite exfoliated GO and rGO (graphene nanosheets in the diagram).

Source: https://pubs.acs.org/doi/10.1021/jp906348x.

1.5.4. Optical Properties of Graphene

Nearly everyone has observed graphene nanosheets accumulated on solid substrates. Indeed, the gray trace remained by the flow of a pencil on a white paper is zero but overlaid graphene nanosheets. Similarly, if pristine graphene nanosheets are scattered in organic solvents, then a gray color is shown by the liquid and becomes darker on the increase in the amount of graphene. As shown in Figure 1.23, a quiet approach to determine the existence of nanoparticles in dispersion is based on the Tyndall effect. A laser beam becomes seeable penetrating to the liquid due to the dispersion of the light by the dispersed nanoparticles.

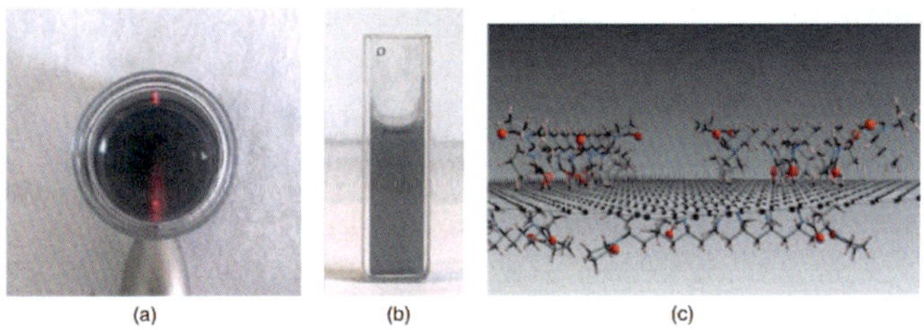

Figure 1.23. (a) A laser beam passing from a dispersion of graphene in water. It is apparent because of the Tyndall scattering effect; (b) the optical transparency of a scattering of graphene in water (0.1 mg ml^{-1}); and (c) diagrammatic model of poly-vinyl pyrrolidone coated graphene.

Source: https://ui.adsabs.harvard.edu/abs/2009SSCom.149.2172B/abstract.

There is a satisfactory light absorption in the graphene as an extended aromatic system; by virtue of an interference effect, if deposited on 300 nm of silicon oxide on top of silicon, even an individual sheet of graphene is apparent from an optical microscope (Jung et al., 2007). Hereafter, various other groups visualize graphene on numerous other substrates. As illustrated in Figure 1.24, graphene's optical absorbance of white light has been evaluated to add up to 2.3%, which means that a five-layer-thick flake absorbs 11.5% and a bilayer absorbs 4.6% (Ni et al., 2008).

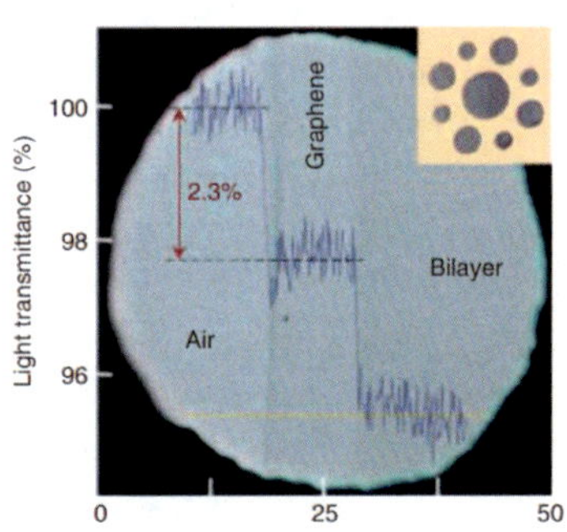

Figure 1.24. A bilayer and single graphene on a porous membrane (Booth et al., 2008).

Source: https://science.sciencemag.org/content/320/5881/1308/tab-article-info.

The upper limit of the absorption is at 268 nm (refers to Figure 1.25(a)). A significantly lower transmittance in comparison with the primary graphite oxide/un-fractioned heparin solution (refers to Figure 1.25(b)) is exhibited by the UV-Vis spectrum of graphene/un-fractioned heparin production after reduction. In DMF, the UV-Vis spectrum of graphene flakes scattered appears very much alike, with a gradually rising curve from 700 to 300 nm (refers to Figure 1.25(c)) (Blake et al., 2007).

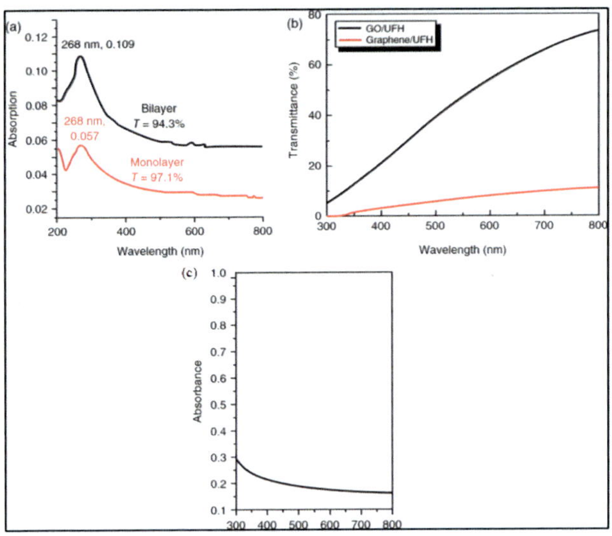

Figure 1.25. (a) The UV-Vis absorption spectra of bilayer graphene and mono-layer graphene; peaks are marked as the value of maximum absorption and the wavelength of maximum absorption. The UV transmittance (T, %) is calculated at 550 nm; (b) UV-Vis spectra of GO and graphene in water solution functional-ized both using heparin (un-fractioned heparin); (c) in DMF, the UV-Vis spectrum of graphene nanosheets

Source: https://pubmed.ncbi.nlm.nih.gov/19408256/.

There is a brownish color of the solid GO and a brownish tint for the dispersions of GO nanosheets as well. When GO is reduced to *r*GO, the color turns to grayer and darker. In contrast to *r*GO or pristine graphene, there is a much higher transmittance of insulated GO because of the various electronic structure (Sun et al., 2010; Lee et al., 2011).

With the reduction of the dimensions of the graphene layers, the electronic characteristics of graphene can be changed as well. As shown in Figure 1.126, there are one or few graphene layers of the graphene quantum dots (GQDs) with size less than 100 nm that exhibit new optoelectronic characteristics, due to the effect of the large percentage of edge atoms and the quantum confinement (Zhang et al., 2012). There is a bandgap in GQDs and shows a strong PL, which can be adjusted by controlling their magnitude and alternate morphological factors (refers to Figure 1.26(C)). At last, with the application of a gate voltage in a field-emitting transistor configuration, the optical transitions in graphene can be changed as well (Wang et al., 2008). This is also an approach to adjust the bandgap in bilayer graphene (Zhang et al., 2009).

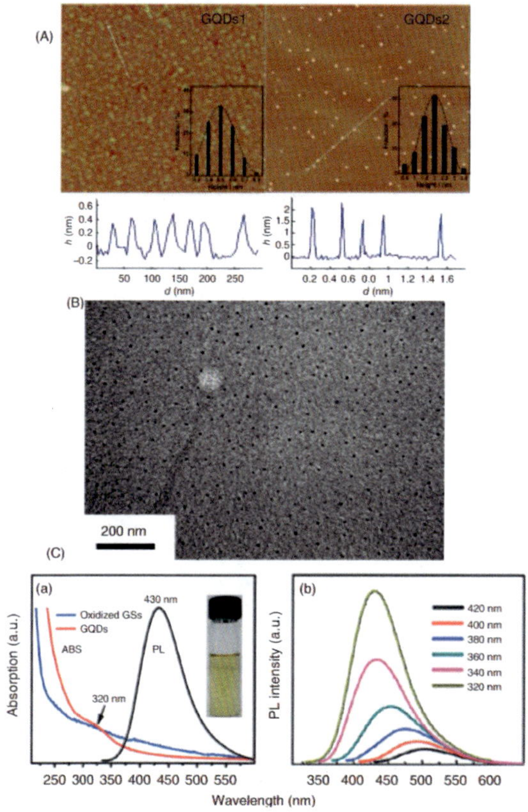

Figure 1.26. (A) and (B) AFM and TEM image of GQDs. (C) (a) UV-Vis absorption (blue line) spectrum of oxidized graphene; UV-Vis absorption (red line) and photolithography (PL) (at 320 nm excitation) spectra of GQDs dispersed in water. Inset: an image of GQD aqueous solution. (b) PL spectra of the GQDs at various excitation wavelengths (Pan et al., 2010).

Source: https://onlinelibrary.wiley.com/doi/abs/10.1002/adma.200902825.

1.5.5. X-Ray Diffraction (XRD) Pattern

Several phases for the isolation of graphene from graphite are appropriately watched by gathering XRD patterns of the initial material, intermediates, and the end product. It is illustrated in Figure 1.27 that a basal reflection (002) peak at $2\theta = 26.6°$ is exhibited by the graphite, which correlated to a *d* spacing of 0.335 nm and indicated the interlayer distance (Yang et al., 2012; Xie et al., 2013).

Before exfoliation and after the oxidation of graphite, the graphite oxide intermediate basal (002) reflection peak is switched to 11.2°, which indicates a *d* spacing of 0.79 nm. The intercalation of water molecules within the oxidized graphene layers connects this rise in the interlayer space. For the verification of the degree of exfoliation, the width of the intense diffraction peak can also be consumed because it is connected to the coherently diffracting domain size from the Debye-Scherrer equation. This diffraction peak vanished on the complete exfoliation of the graphite oxide (Dong et al., 2012; Shen et al., 2012).

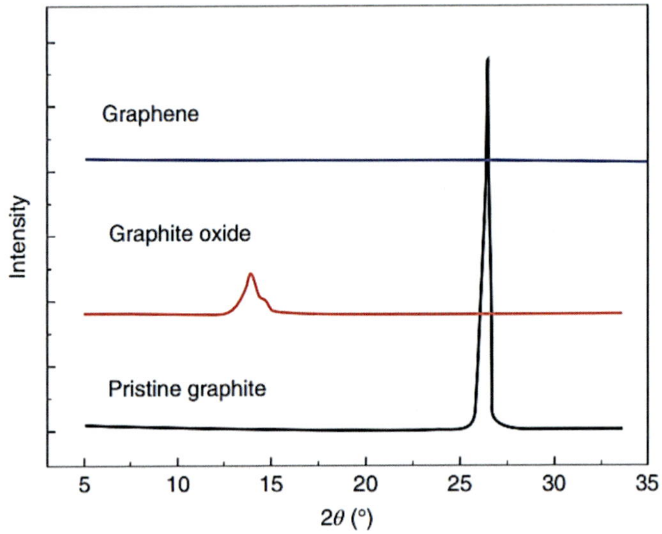

Figure 1.27. X-ray diffraction patterns of graphene, graphite oxide, and pristine graphite (Zhang et al., 2010).

Source: http://www.paper.edu.cn/scholar/showpdf/MUj2YN4IMTD0QxeQh.

REFERENCES

1. Abadal, S., Llatser, I., Mestres, A., Solé-Pareta, J., Alarcón, E., & Cabellos-Aparicio, A., (2017). Fundamentals of graphene-enabled wireless on-chip networking. In: *Modeling, Methodologies and Tools for Molecular and Nano-Scale Communications* (Vol. 1, pp. 293–317). Springer, Cham.
2. Allen, M. J., Tung, V. C., & Kaner, R. B., (2010). Honeycomb carbon: A review of graphene. *Chemical Reviews, 110*(1), 132–145.
3. An, S. J., Zhu, Y., Lee, S. H., Stoller, M. D., Emilsson, T., Park, S., & Ruoff, R. S., (2010). Thin-film fabrication and simultaneous anodic reduction of deposited graphene oxide platelets by electrophoretic deposition. *The Journal of Physical Chemistry Letters, 1*(8), 1259–1263.
4. Batzill, M., (2012). The surface science of graphene: Metal interfaces, CVD synthesis, nanoribbons, chemical modifications, and defects. *Surface Science Reports, 67*(3, 4), 83–115.
5. Blake, P., Brimicombe, P. D., Nair, R. R., Booth, T. J., Jiang, D., Schedin, F., & Geim, A. K., (2008). Graphene-based liquid crystal device. *Nano Letters, 8*(6), 1704–1708.
6. Blake, P., Hill, E. W., Castro, N. A. H., Novoselov, K. S., Jiang, D., Yang, R., & Geim, A. K., (2007). Making graphene visible. *Applied Physics Letters, 91*(6), 063124.
7. Boehm, H. P., Clauss, A., Fischer, G. O., & Hofmann, U., (1962). Thinnest carbon foils. *Zeitschrift Für Naturforschung B, 17*(3), 150–153.
8. Boehm, H. P., Clauss, A., Fischer, G., & Hofmann, U., (1962). Surface properties of extremely thin graphite lamellae. In: *Proceedings of the Fifth Conference on Carbon* (Vol. 1, pp. 73–80). New York: Pergamon Press.
9. Boehm, H. P., Setton, R., & Stumpp, E., (1994). Nomenclature and terminology of graphite intercalation compounds (IUPAC recommendations 1994). *Pure and Applied Chemistry, 66*(9), 1893–1901.
10. Booth, T. J., Blake, P., Nair, R. R., Jiang, D., Hill, E. W., Bangert, U., & Geim, A. K., (2008). Macroscopic graphene membranes and their extraordinary stiffness. *Nano Letters, 8*(8), 2442–2446.

11. Boukhvalov, D. W., & Katsnelson, M. I., (2008). Modeling of graphite oxide. *Journal of the American Chemical Society, 130*(32), 10697–10701.
12. Bourlinos, A. B., Georgakilas, V., Zboril, R., Steriotis, T. A., & Stubos, A. K., (2009a). Liquid-phase exfoliation of graphite towards solubilized graphenes. *Small, 5*(16), 1841–1845.
13. Bourlinos, A. B., Georgakilas, V., Zboril, R., Steriotis, T. A., Stubos, A. K., & Trapalis, C., (2009b). Aqueous-phase exfoliation of graphite in the presence of polyvinylpyrrolidone for the production of water-soluble graphenes. *Solid State Communications, 149*(47, 48), 2172–2176.
14. Bourlinos, A. B., Gournis, D., Petridis, D., Szabó, T., Szeri, A., & Dékány, I., (2003). Graphite oxide: chemical reduction to graphite and surface modification with primary aliphatic amines and amino acids. *Langmuir, 19*(15), 6050–6055.
15. Castro, N. A. H., & Guinea, F., (2009). In: Peres, N. M. R., Novoselov, K. S., & Geim, A. K., (eds.), *Rev. Mod. Phys.* (Vol. 81, pp. 109–162).
16. Chamanara, N., & Caloz, C., (2015). Fundamentals of graphene magneto plasmons: Principles structures and devices. In: *Forum for Electromagn. Res. Methods Appl. Technol.* (Vol. 1, No. 3, p. 3).
17. Chen, D., Li, L., & Guo, L., (2011). An environment-friendly preparation of reduced graphene oxide nanosheets via amino acid. *Nanotechnology, 22*(32), 325601.
18. Chen, W., Yan, L., & Bangal, P. R., (2010). Chemical reduction of graphene oxide to graphene by sulfur-containing compounds. *The Journal of Physical Chemistry C, 114*(47), 19885–19890.
19. Chung, D. D. L., (2002). Review graphite. *Journal of Materials Science, 37*(8), 1475–1489.
20. Coleman, J. N., (2013). Liquid exfoliation of defect-free graphene. *Accounts of Chemical Research, 46*(1), 14–22.
21. Coraux, J., N'Diaye, A. T., Busse, C., & Michely, T., (2008). Structural coherency of graphene on Ir (111). *Nano Letters, 8*(2), 565–570.
22. Cote, L. J., Cruz-Silva, R., & Huang, J., (2009). Flash reduction and patterning of graphite oxide and its polymer composite. *Journal of the American Chemical Society, 131*(31), 11027–11032.

23. Cote, L. J., Kim, F., & Huang, J., (2009). Langmuir-Blodgett assembly of graphite oxide single layers. *Journal of the American Chemical Society, 131*(3), 1043–1049.
24. Diba, M., Fam, D. W., Boccaccini, A. R., & Shaffer, M. S., (2016). Electrophoretic deposition of graphene-related materials: A review of the fundamentals. *Progress in Materials Science, 82*, 83–117.
25. Dong, Y., Chen, C., Zheng, X., Gao, L., Cui, Z., Yang, H., & Li, C. M., (2012). One-step and high yield simultaneous preparation of single- and multi-layer graphene quantum dots from CX-72 carbon black. *Journal of Materials Chemistry, 22*(18), 8764–8766.
26. Dreyer, D. R., Murali, S., Zhu, Y., Ruoff, R. S., & Bielawski, C. W., (2011). Reduction of graphite oxide using alcohols. *Journal of Materials Chemistry, 21*(10), 3443–3447.
27. Emtsev, K. V., Bostwick, A., Horn, K., Jobst, J., Kellogg, G. L., Ley, L., & Rotenberg, E., (2009). Towards wafer-size graphene layers by atmospheric pressure graphitization of silicon carbide. *Nature Materials, 8*(3), 203–207.
28. Fan, X., Peng, W., Li, Y., Li, X., Wang, S., Zhang, G., & Zhang, F., (2008). Deoxygenation of exfoliated graphite oxide under alkaline conditions: A green route to graphene preparation. *Advanced Materials, 20*(23), 4490–4493.
29. Fan, Z. J., Kai, W., Yan, J., Wei, T., Zhi, L. J., Feng, J., & Wei, F., (2011). Facile synthesis of graphene nanosheets via Fe reduction of exfoliated graphite oxide. *ACS Nano, 5*(1), 191–198.
30. Fan, Z., Wang, K., Wei, T., Yan, J., Song, L., & Shao, B., (2010). An environmentally friendly and efficient route for the reduction of graphene oxide by aluminum powder. *Carbon, 48*(5), 1686–1689.
31. Fernández-Merino, M. J., Guardia, L., Paredes, J. I., Villar-Rodil, S., Solís-Fernández, P., Martínez-Alonso, A., & Tascón, J. M. D., (2010). Vitamin C is an ideal substitute for hydrazine in the reduction of graphene oxide suspensions. *The Journal of Physical Chemistry C,114*(14), 6426–6432.
32. Ferrari, A. C., Meyer, J. C., Scardaci, V., Casiraghi, C., Lazzeri, M., Mauri, F., & Geim, A. K., (2006). Raman spectrum of graphene and graphene layers. *Physical Review Letters, 97*(18), 187401.
33. Frank, I. W., Tanenbaum, D. M., Van, D. Z. A. M., & McEuen, P. L., (2007). Mechanical properties of suspended graphene sheets. *Journal of Vacuum Science and Technology B: Microelectronics and Nanometer*

Structures Processing, Measurement, and Phenomena, 25(6), 2558–2561.

34. Gao, W., Alemany, L. B., Ci, L., & Ajayan, P. M., (2009). New insights into the structure and reduction of graphite oxide. *Nature Chemistry, 1*(5), 403–408.
35. Gengler, R. Y., Veligura, A., Enotiadis, A., Diamanti, E. K., Gournis, D., Józsa, C., & Rudolf, P., (2010). Large-yield preparation of high-electronic-quality graphene by a Langmuir-Schaefer approach. *Small, 6*(1), 35–39.
36. Gómez-Navarro, C., Weitz, R. T., Bittner, A. M., Scolari, M., Mews, A., Burghard, M., & Kern, K., (2007). Electronic transport properties of individual chemically reduced graphene oxide sheets. *Nano Letters, 7*(11), 3499–3503.
37. Guittonneau, F., Abdelouas, A., Grambow, B., & Huclier, S., (2010). The effect of high-power ultrasound on an aqueous suspension of graphite. *Ultrasonics Sonochemistry, 17*(2), 391–398.
38. Hamilton, C. E., Lomeda, J. R., Sun, Z., Tour, J. M., & Barron, A. R., (2009). High-yield organic dispersions of unfunctionalized graphene. *Nano Letters, 9*(10), 3460–3462.
39. Han, M. Y., Özyilmaz, B., Zhang, Y., & Kim, P., (2007). Energy band-gap engineering of graphene nanoribbons. *Physical Review Letters, 98*(20), 206805.
40. He, H., Klinowski, J., Forster, M., & Lerf, A., (1998). A new structural model for graphite oxide. *Chemical Physics Letters, 287*(1, 2), 53–56.
41. Hernandez, Y., Nicolosi, V., Lotya, M., Blighe, F. M., Sun, Z., De, S., & Boland, J. J., (2008). High-yield production of graphene by liquid-phase exfoliation of graphite. *Nature Nanotechnology, 3*(9), 563–568.
42. Huang, P. Y., Ruiz-Vargas, C. S., Van, D. Z. A. M., Whitney, W. S., Levendorf, M. P., Kevek, J. W., & Park, J., (2011). Grains and grain boundaries in single-layer graphene atomic patchwork quilts. *Nature, 469*(7330), 389–392.
43. Hummers, Jr. W. S., & Offeman, R. E., (1958). Preparation of graphitic oxide. *Journal of the American Chemical Society, 80*(6), 1339–1339.
44. Jiao, L., Wang, X., Diankov, G., Wang, H., & Dai, H., (2010). Facile synthesis of high-quality graphene nanoribbons. *Nature Nanotechnology, 5*(5), 321–325.

45. Jinschek, J. R., Yucelen, E., Calderon, H. A., & Freitag, B., (2011). Quantitative atomic 3-D imaging of single/double sheet graphene structure. *Carbon, 49*(2), 556–562.

46. Jung, I., Pelton, M., Piner, R., Dikin, D. A., Stankovich, S., Watcharotone, S., & Ruoff, R. S., (2007). Simple approach for high-contrast optical imaging and characterization of graphene-based sheets. *Nano Letters, 7*(12), 3569–3575.

47. Khan, U., O'Neill, A., Lotya, M., De, S., & Coleman, J. N., (2010). High-concentration solvent exfoliation of graphene. *Small, 6*(7), 864–871.

48. Kim, J., Jeon, J. H., Kim, H. J., Lim, H., & Oh, I. K., (2014). Durable and water-floatable ionic polymer actuator with hydrophobic and asymmetrically laser-scribed reduced graphene oxide paper electrodes. *ACS Nano, 8*(3), 2986–2997.

49. Kim, K. S., Zhao, Y., Jang, H., Lee, S. Y., Kim, J. M., Kim, K. S., & Hong, B. H., (2009). Large-scale pattern growth of graphene films for stretchable transparent electrodes. *Nature, 457*(7230), 706–710.

50. Kim, O., Shin, T. J., & Park, M. J., (2013). Fast low-voltage electroactive actuators using nanostructured polymer electrolytes. *Nature Communications, 4*(1), 1–9.

51. Kong, L., & Chen, W., (2014). Carbon nanotube and graphene-based bioinspired electrochemical actuators. *Advanced Materials, 26*(7), 1025–1043.

52. Kosidlo, U., Omastová, M., Micusík, M., Ćirić-Marjanović, G., Randriamahazaka, H., Wallmersperger, T., & Bauernhansl, T., (2013). Nanocarbon based ionic actuators: A review. *Smart Materials and Structures, 22*(10), 104022.

53. Kotal, M., Kim, J., Kim, K. J., & Oh, I. K., (2016). Sulfur and nitrogen co-doped graphene electrodes for high-performance ionic artificial muscles. *Advanced Materials, 28*(8), 1610–1615.

54. Lee, D. Y., Khatun, Z., Lee, J. H., Lee, Y. K., & In, I., (2011). Blood compatible graphene/heparin conjugate through noncovalent chemistry. *Biomacromolecules, 12*(2), 336–341.

55. Lee, J. H., Lee, E. K., Joo, W. J., Jang, Y., Kim, B. S., Lim, J. Y., & Yang, C. W., (2014). Wafer-scale growth of single-crystal monolayer graphene on reusable hydrogen-terminated germanium. *Science, 344*(6181), 286–289.

56. Leon, V., Quintana, M., Herrero, M. A., Fierro, J. L., De La Hoz, A., Prato, M., & Vazquez, E., (2011). Few-layer graphenes from ball-milling of graphite with melamine. *Chemical Communications, 47*(39), 10936–10938.

57. Li, J., Ma, W., Song, L., Niu, Z., Cai, L., Zeng, Q., & Xie, S., (2011). Superfast-response and ultrahigh-power-density electromechanical actuators based on hierarchal carbon nanotube electrodes and chitosan. *Nano Letters, 11*(11), 4636–4641.

58. Liu, S., Tian, J., Wang, L., & Sun, X., (2011). A method for the production of reduced graphene oxide using benzylamine as a reducing and stabilizing agent and its subsequent decoration with Ag nanoparticles for enzyme less hydrogen peroxide detection. *Carbon, 49*(10), 3158–3164.

59. Liu, Y., Dai, H., Deng, J., Zhang, L., Zhao, Z., Li, X., & Guo, G., (2013). Controlled generation of uniform spherical $LaMnO_3$, $LaCoO_3$, Mn_2O_3, and Co_3O_4 nanoparticles and their high catalytic performance for carbon monoxide and toluene oxidation. *Inorganic Chemistry, 52*(15), 8665–8676.

60. Lotya, M., King, P. J., Khan, U., De, S., & Coleman, J. N., (2010). High-concentration, surfactant-stabilized graphene dispersions. *ACS Nano, 4*(6), 3155–3162.

61. Lu, L., Liu, J., Hu, Y., Zhang, Y., & Chen, W., (2013). Graphene-stabilized silver nanoparticle electrochemical electrode for actuator design. *Advanced Materials, 25*(9), 1270–1274.

62. Lu, L., Liu, J., Hu, Y., Zhang, Y., Randriamahazaka, H., & Chen, W., (2012). Highly stable air working bimorph actuator based on a graphene nanosheet/carbon nanotube hybrid electrode. *Advanced Materials, 24*(31), 4317–4321.

63. Mattevi, C., Kim, H., & Chhowalla, M., (2011). A review of chemical vapor deposition of graphene on copper. *Journal of Materials Chemistry, 21*(10), 3324–3334.

64. Meng, L., Sun, Q., Wang, J., & Ding, F., (2012). Molecular dynamics simulation of chemical vapor deposition graphene growth on Ni (111) surface. *The Journal of Physical Chemistry C,116*(10), 6097–6102.

65. Meyer, J. C., Geim, A. K., Katsnelson, M. I., Novoselov, K. S., Booth, T. J., & Roth, S., (2007). The structure of suspended graphene sheets. *Nature, 446*(7131), 60–63.

66. Misik, V., & Riesz, P., (1996). Peroxyl radical formation in aqueous solutions of N, N-dimethylformamide, N-methylformamide, and dimethylsulfoxide by ultrasound: Implications for sonosensitized cell killing. *Free radical Biology and Medicine, 20*(1), 129.
67. Mittal, V., (2012). *Polymer-Graphene Nanocomposites* (Vol. 26, pp. 1–33). Royal Society of Chemistry.
68. Mohiuddin, T. M. G., Lombardo, A., Nair, R. R., Bonetti, A., Savini, G., Jalil, R., & Novoselov, K. S., (2009). Uniaxial strain in graphene by Raman spectroscopy: G peak splitting, Grüneisen parameters, and sample orientation. *Physical Review B, 79*(20), 205433.
69. Moon, I. K., Lee, J., Ruoff, R. S., & Lee, H., (2010). Reduced graphene oxide by chemical graphitization. *Nature Communications, 1*(1), 1–6.
70. Moon, J. S., & Gaskill, D. K., (2011). Graphene: Its fundamentals to future applications. *IEEE Transactions on Microwave Theory and Techniques, 59*(10), 2702–2708.
71. Morales-Torres, S., Pastrana-Martínez, L. M., Figueiredo, J. L., Faria, J. L., & Silva, A. M., (2012). Design of graphene-based TiO_2 photocatalysts: A review. *Environmental Science and Pollution Research, 19*(9), 3676–3687.
72. Mukai, K., Asaka, K., Sugino, T., Kiyohara, K., Takeuchi, I., Terasawa, N., & Aida, T., (2009). Highly conductive sheets from millimeter-long single-walled carbon nanotubes and ionic liquids: Application to fast-moving, low-voltage electromechanical actuators operable in air. *Advanced Materials, 21*(16), 1582–1585.
73. Ni, Z. H., Chen, W., Fan, X. F., Kuo, J. L., Yu, T., Wee, A. T. S., & Shen, Z. X., (2008). Raman spectroscopy of epitaxial graphene on a SiC substrate. *Physical Review B,77*(11), 115416.
74. Novoselov, K. S., Fal, V. I., Colombo, L., Gellert, P. R., Schwab, M. G., & Kim, K., (2012). A roadmap for graphene. *Nature, 490*(7419), 192–200.
75. Novoselov, K. S., Geim, A. K., Morozov, S. V., Jiang, D., Katsnelson, M. I., Grigorieva, I., & Firsov, A. A., (2005). Two-dimensional gas of massless Dirac fermions in graphene. *Nature, 438*(7065), 197–200.
76. Novoselov, K. S., Geim, A. K., Morozov, S. V., Jiang, D., Zhang, Y., Dubonos, S. V., & Firsov, A. A., (2004). Electric field effect in atomically thin carbon films. *Science, 306*(5696), 666–669.

77. Obeng, Y., & Srinivasan, P., (2011). Graphene: Is it the future for semiconductors? An overview of the material, devices, and applications. *Electrochemical Society Interface, 20*(1), 47.
78. Pan, D., Zhang, J., Li, Z., & Wu, M., (2010). Hydrothermal route for cutting graphene sheets into blue-luminescent graphene quantum dots. *Advanced Materials, 22*(6), 734–738.
79. Park, J. S., Reina, A., Saito, R., Kong, J., Dresselhaus, G., & Dresselhaus, M. S., (2009). G′ band Raman spectra of single, double, and triple layer graphene. *Carbon, 47*(5), 1303–1310.
80. Pei, S., Zhao, J., Du, J., Ren, W., & Cheng, H. M., (2010). Direct reduction of graphene oxide films into highly conductive and flexible graphene films by hydrohalic acids. *Carbon, 48*(15), 4466–4474.
81. Perrozzi, F., Prezioso, S., & Ottaviano, L., (2014). Graphene oxide: From fundamentals to applications. *Journal of Physics: Condensed Matter, 27*(1), 013002.
82. Pop, E., Varshney, V., & Roy, A. K., (2012). Thermal properties of graphene: Fundamentals and applications. *MRS Bulletin, 37*(12), 1273–1281.
83. Prezhdo, O. V., (2011). Graphene: The ultimate surface material. *Sur. Sc., 605*(17), 1607–1610.
84. Quintana, M., Grzelczak, M., Spyrou, K., Kooi, B., Bals, S., Van, T. G., & Prato, M., (2012). Production of large graphene sheets by exfoliation of graphite under high power ultrasound in the presence of tiopronin. *Chemical Communications, 48*(100), 12159–12161.
85. Rana, K., Singh, J., & Ahn, J. H., (2014). A graphene-based transparent electrode for use in flexible optoelectronic devices. *Journal of Materials Chemistry C, 2*(15), 2646–2656.
86. Reina, A., Jia, X., Ho, J., Nezich, D., Son, H., Bulovic, V., & Kong, J., (2009). Large area, few-layer graphene films on arbitrary substrates by chemical vapor deposition. *Nano Letters, 9*(1), 30–35.
87. Schniepp, H. C., Li, J. L., McAllister, M. J., Sai, H., Herrera-Alonso, M., Adamson, D. H., & Aksay, I. A., (2006). Functionalized single graphene sheets derived from splitting graphite oxide. *The Journal of Physical Chemistry B, 110*(17), 8535–8539.
88. Shen, J., Zhu, Y., Yang, X., Zong, J., Zhang, J., & Li, C., (2012). One-pot hydrothermal synthesis of graphene quantum dots surface-passivated by polyethylene glycol and their photoelectric conversion

under near-infrared light. *New Journal of Chemistry, 36*(1), 97–101.

89. Shin, H. J., Kim, K. K., Benayad, A., Yoon, S. M., Park, H. K., Jung, I. S., & Lee, Y. H., (2009). Efficient reduction of graphite oxide by sodium borohydride and its effect on electrical conductance. *Advanced Functional Materials, 19*(12), 1987–1992.
90. Somani, P. R., Somani, S. P., & Umeno, M., (2006). Planer nano-graphenes from camphor by CVD. *Chemical Physics Letters, 430*(1–3), 56–59.
91. Sounas, D. L., & Caloz, C., (2012). Gyrotropy and nonreciprocity of graphene for microwave applications. *IEEE Transactions on Microwave Theory and Techniques, 60*(4), 901–914.
92. Stankovich, S., Dikin, D. A., Dommett, G. H., Kohlhaas, K. M., Zimney, E. J., Stach, E. A., & Ruoff, R. S., (2006). Graphene-based composite materials. *Nature, 442*(7100), 282–286.
93. Stankovich, S., Dikin, D. A., Piner, R. D., Kohlhaas, K. A., Kleinhammes, A., Jia, Y., & Ruoff, R. S., (2007). Synthesis of graphene-based nanosheets via chemical reduction of exfoliated graphite oxide. *Carbon, 45*(7), 1558–1565.
94. Staudenmaier, L., (1898). Process for the preparation of graphitic acid. *Ber. Dtsch. Chem. Ges., 31*, 1481–1487.
95. Sun, Z., Yan, Z., Yao, J., Beitler, E., Zhu, Y., & Tour, J. M., (2010). Growth of graphene from solid carbon sources. *Nature, 468*(7323), 549–552.
96. Suslick, K. S., (1990). Sonochemistry. *Science, 247*(4949), 1439–1445.
97. Sutter, P. W., Flege, J. I., & Sutter, E. A., (2008). Epitaxial graphene on ruthenium. *Nature Materials, 7*(5), 406–411.
98. Terrones, M., Botello-Méndez, A. R., Campos-Delgado, J., López-Urías, F., Vega-Cantú, Y. I., Rodríguez-Macías, F. J., & Terrones, H., (2010). Graphene and graphite nanoribbons: Morphology, properties, synthesis, defects, and applications. *Nano Today, 5*(4), 351–372.
99. Tian, H., Shu, Y., Cui, Y. L., Mi, W. T., Yang, Y., Xie, D., & Ren, T. L., (2014). Scalable fabrication of high-performance and flexible graphene strain sensors. *Nanoscale, 6*(2), 699–705.
100. Tombros, N., Jozsa, C., Popinciuc, M., Jonkman, H. T., & Van, W. B. J., (2007). Electronic spin transport and spin precession in single graphene layers at room temperature. *Nature, 448*(7153), 571–574.

101. Tuinstra, F., & Koenig, J. L., (1970). Raman spectrum of graphite. *The Journal of Chemical Physics, 53*(3), 1126–1130.

102. Wang, F., Zhang, Y., Tian, C., Girit, C., Zettl, A., Crommie, M., & Shen, Y. R., (2008). Gate-variable optical transitions in graphene. *Science, 320*(5873), 206–209.

103. Wang, G., Yang, J., Park, J., Gou, X., Wang, B., Liu, H., & Yao, J., (2008). Facile synthesis and characterization of graphene nanosheets. *The Journal of Physical Chemistry C,112*(22), 8192–8195.

104. Wang, X., & Shi, G., (2015). Flexible graphene devices related to energy conversion and storage. *Energy and EnvironmentalScience, 8*(3), 790–823.

105. Wei, D., Wu, B., Guo, Y., Yu, G., & Liu, Y., (2013). Controllable chemical vapor deposition growth of few-layer graphene for electronic devices. *Accounts of Chemical Research, 46*(1), 106–115.

106. Wu, G., Hu, Y., Liu, Y., Zhao, J., Chen, X., Whoehling, V., & Chen, W., (2015). Graphitic carbon nitride nanosheet electrode-based high-performance ionic actuator. *Nature Communications, 6*(1), 1–8.

107. Wu, S., He, Q., Tan, C., Wang, Y., & Zhang, H., (2013). Graphene-based electrochemical sensors. *Small, 9*(8), 1160–1172.

108. Xie, M., Su, Y., Lu, X., Zhang, Y., Yang, Z., & Zhang, Y., (2013). Blue and green photoluminescence graphene quantum dots synthesized from carbon fibers. *Materials Letters, 93*, 161–164.

109. Xie, X., Kretschmer, K., & Wang, G., (2015). Advances in graphene-based semiconductor photocatalysts for solar energy conversion: Fundamentals and materials engineering. *Nanoscale, 7*(32), 13278–13292.

110. Yang, F., Zhao, M., Zheng, B., Xiao, D., Wu, L., & Guo, Y., (2012). Influence of pH on the fluorescence properties of graphene quantum dots using ozonation pre-oxide hydrothermal synthesis. *Journal of Materials Chemistry, 22*(48), 25471–25479.

111. Zhang, H. B., Zheng, W. G., Yan, Q., Yang, Y., Wang, J. W., Lu, Z. H., & Yu, Z. Z., (2010). Electrically conductive polyethylene terephthalate/graphene nanocomposites prepared by melt compounding. *Polymer, 51*(5), 1191–1196.

112. Zhang, M., Bai, L., Shang, W., Xie, W., Ma, H., Fu, Y., & Liu, C., (2012). Facile synthesis of water-soluble, highly fluorescent graphene

quantum dots as a robust biological label for stem cells. *Journal of Materials Chemistry, 22*(15), 7461–7467.

113. Zhang, Y., Tang, T. T., Girit, C., Hao, Z., Martin, M. C., Zettl, A., & Wang, F., (2009). Direct observation of a widely tunable bandgap in bilayer graphene. *Nature, 459*(7248), 820–823.
114. Zhou, G., Li, F., & Cheng, H. M., (2014). Progress in flexible lithium batteries and future prospects. *Energy and Environmental Science, 7*(4), 1307–1338.
115. Zhou, M., Wang, Y., Zhai, Y., Zhai, J., Ren, W., Wang, F., & Dong, S., (2009). Controlled synthesis of large-area and patterned electrochemically reduced graphene oxide films. *Chemistry-a European Journal, 15*(25), 6116–6120.
116. Zhou, T., Chen, F., Liu, K., Deng, H., Zhang, Q., Feng, J., & Fu, Q., (2010). A simple and efficient method to prepare graphene by reduction of graphite oxide with sodium hydrosulfite. *Nanotechnology, 22*(4), 045704.
117. Zhou, X., Zhang, J., Wu, H., Yang, H., Zhang, J., & Guo, S., (2011). Reducing graphene oxide via hydroxylamine: A simple and efficient route to graphene. *The Journal of Physical Chemistry C,115*(24), 11957–11961.

CHAPTER 2

Synthesis Methods of Graphene and Its Properties

CONTENTS

2.1. INTRODUCTION

The age of stone, iron, and copper is now ancient history, currently, carbon, steel, and nearly perfect silicone are the materials of choice. Yet, carbon is the one that still got a lot of consideration. Identified as a nonrenewable energy source, carbon is an omnipresent molecule capable of developing many allotropes with numerous potential applications. The main ones are diamond and graphite (Falcao and Wudl, 2007; Segal, 2012). Lately, carbon had become again an object of extreme scientific research which led to numerous discoveries. In 1996 Smalley and his colleagues got the Nobel Prize in Chemistry "due to their discovery of fullerenes," zero-dimensional nanospheres made by 60 carbon atoms which look like the soccer ball and occur only in the molecular form of carbon in comparison to the crystalline forms of the diamond and graphite. In 1991, Iijima described one-dimensional helical microtubules of graphitic carbon termed nanotubes (Iijima, 1991; Smalley, 1997). Once more in 2010, the Nobel Prize in Physics was awarded jointly to Konstatntin Novoselov and Andre Geim "for revolutionary experiments about the two-dimensional (2D) material graphene" (Novoselov et al., 2004). The scientists got 2D graphene merely one atom thick, amazingly reminiscent of a honeycomb structure through the simple usage of normal sticky tape. They not only make graphene however they also studied and explained its possible applications and unique properties (Geim, 2009, 2011). Graphene is a key element for the other graphitic materials, and it is probable to convert one structure to another under suitable conditions (Figure 2.1). The previous 20 years of research had shown numerous possibilities of chemical functionalization of diverse synthetic carbon diversities like carbon nanotubes (CNTs) or fullerenes which led to numerous significant attainments in improving the solubility, fixing with other compounds, processing capabilities, and discovering the unique properties of the graphene (Novoselov, 2011; Yan et al., 2011). Diverse dispersal of the benzene rings in the 2D structure of graphene concludes the size, shape, the number of layers, edges, and additional non-covalent or covalent bonds with other atoms, which consequences in modifications of the chemical and electrical properties of graphene. Like in graphene and semiconductors, the electrical current is transmitted either through negatively charged electrons or through the positively charged holes that were left behind. Though, on differing to the conventional semiconductors the 2D graphene layer demonstrates a strong ambipolar electric field influence in the room temperature with the least band gap amongst the conduction bands (CBs) and the valence. It permits ballistic electron to transmit over long distances and with a high speed that is merely 300 times slower than the speed of light and from 10 to 100 times superior to that in the presently used silicon chips. Graphene is transparent, flexible, and well-conductive material with outstanding mechanical properties and is the thinnest identified structure being 200 times larger than steel and harder than diamond (Allen et al., 2010; Fuhrer et al., 2010). The omnipresent and extensively available carbon is the basic unit of graphene and it is not astonishing that it had become a growing star in the current world of nanomaterials, although being competitive for silicone, a necessary material of recent electronics. To allow the usage of graphene and its distinctive properties and to assure economical production and its application in the industry it would be essential to bypass some main hindrances. Those are high costs and problems in the production of graphene on a huge scale and the qualitative and quantitative control of the subsequent product (Singh et al., 2011; Eigler and Hirsch, 2014). Furthermore, each of the graphene production techniques results in various kinds of the same material, which had a different quality, properties, number, and kinds of defects. Collecting evidence from laboratories regarding rare properties of graphene had been carried out on high quality, little, and single-layer samples without contaminations or structural defects, which might not essentially be reflected in graphene got on a huge scale. Producing ideal crystals formed of a single layer at low cost and specific manipulation on the micro-level is yet a big challenge regardless of the numerous known techniques of synthesis and graphene production (Zhu et al., 2010; Van Noorden, 2012). Researchers from diverse fields of science are attentive to graphene and are still knowing how to work with it and taking benefit of its infinite potential, which could be confirmed by an increasing number of publications regarding its latest applications (Edwards and Coleman, 2013; Eremina et al., 2018).

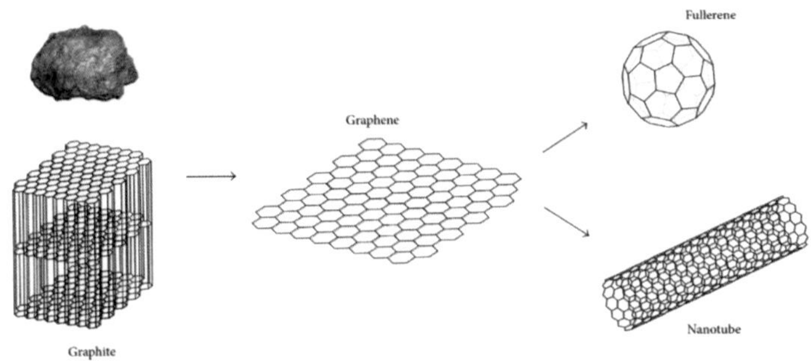

Figure 2.1. Graphite is a usual and naturally found mineral in nature. Graphene is a basic element for the further graphitic materials, and it is probable to transform one structure to another in suitable conditions.

Source: Source: https://www.hindawi.com/journals/jnm/2014/890246/.

2.2. THE PROPERTIES AND APPLICATIONS OF GRAPHENE

Graphene is a single layer of atomic carbon, and its theoretic thickness is only 0.34 nanometer. The theoretical precise surface area of graphene is up to 2600 m^2/g. It has exceptional thermal conductivity (3000 W/(m K)), also has high-speed electron mobility (15000 $cm^2/(V\ s)$) at room temperature. Its density is only 2.2 g/cm^3 while the mechanical stress reaches 1060 GPa (Soldano et al., 2010; Mao et al., 2013).

Moreover, graphene also had numerous special characteristics, like chiral tunneling of relativistic particles, lack of Anderson location, bipolar supercurrent, and anomalous quantum Hall impact (Wei and Liu, 2010). These distinctive characters make it gives a research platform for quantum mechanics and compressed matter physics. Numerous phenomena are easily perceived in graphene, which was only perceived in heavy-ion accelerators or black holes in previous (Bitounis et al., 2013; Zhang et al., 2013). Furthermore, graphene is the toughest material ever measured on earth. It is noted that the graphene density is even tougher than diamond and around 100 times stronger than that of the finest steels in the world.

The single graphene layer is a zero-gap or semimetal semiconductor and has outstanding electronic properties. The electron mobility of graphene is very great; it is around 10 times quicker than the commercial silicon wafer. Moreover, mobility is slightly affected by the temperature. Its carrier demonstrates ballistic transport under an ambient situation (Boehm et al., 1994).

These properties share prominent benefits of graphene applied in nanoelectronics devices. It is probable using graphene for ballistic field impact transistor at room temperature. The nanoribbon transistors with huge on-off current ratios at room temperature were displayed in Figure 2.2(b). Graphene makes it probable for this comparatively simple framework to execute complex and novel circuit and becomes the basic electronic material beside the silicon age (Ferrari et al., 2006).

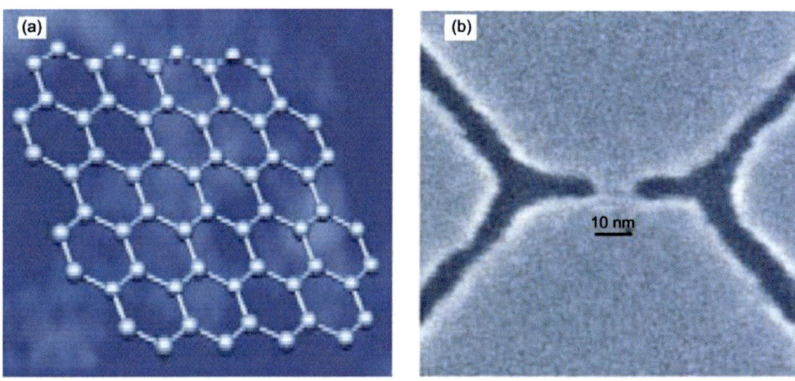

Figure 2.2. (a) The structure of graphene; (b) graphene nanoribbons show the transistor action with huge on-off ratios.

Source: https://pubmed.ncbi.nlm.nih.gov/18420930/.

Though the present situation is not optimistic, and the complications are still puzzling researchers, like the preparation of huge area high-quality graphene, the high constancy, and consistency, as well as a simple procedure. To achieve the size required, we frequently choose to cut down graphene to get graphene nanoribbons (GNRs). GNRs had already been formed employing STM (scanning tunneling microscopy) lithography, plasma etching, and AFM (atomic force microscopy) anodic oxidation, also as through chemically derived methods. Though, a method to form single-layer graphene structures by well-defined crystallographic edges remains obscure. Geim (2009) predicted that it might need another 10 years earlier for the first integrated circuits (ICs) founded on graphene chips to look.

Another significant application is that graphene could be used as a transparent electrode, which might be widely employed in LCD (liquid crystal display), OLED (organic light-emitting display), or organic solar cells are yet to come (Chia et al., 2013). Graphene had an optical transmission rate above 98%, significantly higher than that 82%–85% of the standard ITO (indium tin oxides) film. However, the reduction processing graphene exhibits a low resistance, and it could be minimized furthermore through chemical doping. On the other hand, ITO has to trade resistance for clarity. If an ITO electrode is prepared thin enough to compete with the transparency of graphene, its sheet opposition should also be skyrockets.

Established on a huge precise surface area, graphene is also utilized as a biological system and gas detector. Schedin et al. (2007) displayed that the micrometer-sized sensors formed from graphene are capable of identifying individual events when a gas molecule is detached or attached from graphene's surface. Mohanty et al. (2012) revealed the interfacing of chemically altered graphene with biological systems to build a DNA-hybridization device and a new live-bacterial-hybrid device with outstanding sensitivity (Obst et al., 2017).

Due to the outstanding thermal, mechanical, and electrical properties, graphene is predicted to be employed in high-performance nano-electronic devices, energy storage areas, field emission (FE) materials, gas sensors, composite materials, etc.

2.3. SYNTHESIS

In 2004 the discovery of graphene was not the first interaction with this revolutionary material. Previously in 1986, Boehm and coworkers noticed monolayer graphene and officially termed it using a grouping of the word graphite and the suffix "en" linking to polycyclic aromatic hydrocarbons. It was the development of the new methods which permitted to identify and describe the single-atom structure of graphene and its scarce properties, though. Graphene is perhaps produced all time while using a pencil. Though, it is difficult to perceive it amongst other stacked layers of graphite. In actuality, the observation of graphene is not probable using traditional visualization methods (with the exclusion of atomic force or electron microscopy) due to the absence of the clear observable differences in the atomic structure of graphene monolayer and numerous layers of graphite (Park and Ruoff, 2009). To get the wanted image intensity and contrast and to recognize the single atom structure of graphene in the optical microscope, it is essential to select the suitable substrate of specific wavelength and thickness. Excluding the optical microscopy, there are methods beneficial in observing and determining the structure of graphenes like AFM, APRES (angular distribution photoemission spectroscopy), TEM (transmission electron microscopy), and the spectroscopy of Raman and Rayleigh. Raman spectroscopy is one of the best techniques for insight into the system of graphene, quick, and precise characterization, and direct measurements of the electron interactions, which permit determining the number of layers and identify flaws and impurities. The Raman spectrum describes three major bands: the band D~ 1350 cm^{-1}, G band~ 1580 cm^{-1}, activated through defects, and band 2D ~2700 cm^{-1} which tells about the number of layers and the value of the cumulative load. The rise in the ratio of the D peak concentration to that of the G peak and expansion of both of them is caused by an enhanced number of defects on the surface of graphene. The Raman spectroscopy mainly gives information regarding the consistency of the graphene carbon skeleton. In turn, the concentration of D peak rises slowly together with the number of holes in the structure or novel centers with sp^3 hybridization formed by covalent bonds (Figure 2.3). Though, it is hard to differentiate between them. In the existence of less than 1% of the structural flaws, the analysis of the D band could be utilized to decide the quality of graphene got by minimizing the graphene oxide (GO) or to state the degree of the functionalization.

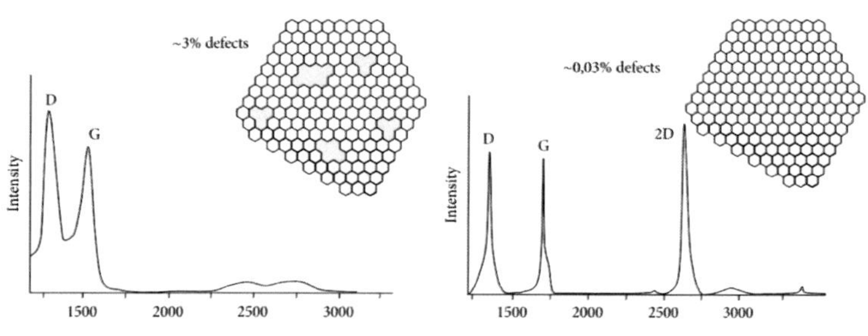

Figure 2.3. Raman spectra of graphene with simplified structural models. (a) GO with flaw density of 1–3%; (b) GO with nearly intact carbon framework with 0.03% flaws.

Source: https://www.researchgate.net/publication/221703598_Production_Beyond_sticky_tape.

Konstantin Novoselov and Andre Geim used an effective and simple technique to get graphene by the repeated utilization of the adhesive tape to rip off tinny flakes from the graphite and then attached them to a plate of SiO_2 oxidized silicon to disclose and determine the number of graphene layers.

Primarily due to the utilization of high-quality crystals of graphite (HOPG: highly ordered pyrolytic graphite) as an initiating material the mechanical exfoliation yet remains one of the best techniques to obtain electrically and structurally homogenous and single-layer graphene. Regardless of being limited by its little production and the probability of practical use, the mechanical exfoliation technique led to many discoveries of graphene mechanical and electronic properties and resulted in the progress of new techniques of production.

There are two methods to obtain graphene (Figure 2.4 and Table 2.1). One comprises the graphite and weakening of the van der Waals forces to distinct the layers from each other, and the second is founded on alternate carbon sources. The effective partition without harming the structure and protection beside reaggregation are the main elements to get graphene from graphite. The drawbacks of these techniques are low yield, various steps of the production, and the point that natural graphite is not simply available and it is registered in the European List of Critical Raw Materials.

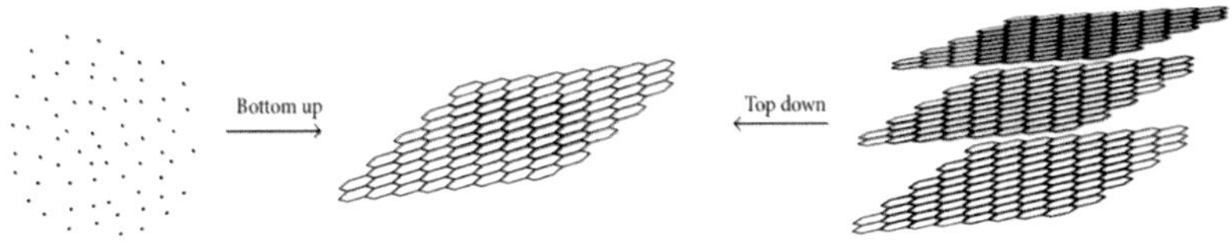

Figure 2.4. A diagram of bottom-up and top-down methods for graphene synthesis.

Source: https://pubs.rsc.org/en/content/articlelanding/2013/nr/c2nr32629a.

In turn, graphite synthetic form got at high temperatures is not appropriate for the production of high-quality graphene due to its irregular structure. Noncovalent interactions amongst every layer in natural graphite make an exceptionally stable and thermodynamically satisfactory arrangement which is the cause of the reaggregation of graphene flakes, for instance, during graphite exfoliation in solution using chemical techniques (Li et al., 2009).

In assumption, the techniques of production of graphene de novo are very easier and permit getting huge area surfaces on chosen substrates however require high temperatures and produce structure damage and lots of defects. These techniques comprise epitaxial growth on a solid substrate and chemical vapor deposition (CVD) which needs strict control, specialized equipment, and laborious preparation methods (Meca et al., 2013; Yang et al., 2013). The epitaxial growth technique originally includes putting silicon carbide (SiC) in a vacuum at a temperature of 1300°C which consequence in the sublimation of Si atoms and restructuring of the residual carbon atoms. The temperature control and exact time of the procedure should lead to the creation of thin layers of graphene on the whole surface of SiC wafers which rarely forms monolayer. Though, the comprehensive studies of these structures display very weak bonding and surface roughness, rotation of the individual layers, and the presence of several cavities and holes. Over time and to get more firm and homogenous graphene monolayers the novel developments have been introduced like the change of atmosphere and the temperature of the sublimation procedure (Obraztsov et al., 2007).

The techniques are based on the dispersal of hydrocarbons, and the deposition of the graphene on the selected substrate has a lot of interest due to its earlier application in CNTs production (Bae et al., 2010). The CVD on the metallic surfaces which catalyzes the sustained growth of graphene needs specific conditions. In the initial method, the procedure was carried out in the ethylene atmosphere at 800 K temperature with Pt as a surface substrate. To develop the procedure, the type of the gas (for example, benzene), the temperature, and metal surface (e.g., Ni, Cu, Ir, Ru, Ir, Ni, or Cu) were altered. The mixture of the terms of the chemisorption of the graphene on the metal surfaces permits obtaining a homogenous and regular graphene layer which could be transmitted later on to any kind of material. The development of the monolayer graphene was also attained through the thermal decomposition of small molecules or polymer films and even from unconventional carbon sources like chocolate, grass, cockroach legs, or biscuits (Wei et al., 2007; Sun et al., 2010).

The techniques of supported growth of graphene need costly templates which would be beneficial in high performance and more demanding applications. Very much striking from an economic point of view is the chemical techniques for the manufacturing of graphene in the solution (Ruan et al., 2011; Chia et al., 2013). They give the probability of mass production with the suitable optical and electrical properties of graphene. Though, during the making of the graphene from GO the firm control of the procedure conditions is needed to prevent extreme oxidation, the release of CO_2, and the creation of impossible to eradicate lattice defects. GO is an extremely oxidized form of chemically altered graphene produced through the oxidation of crystalline graphite in solution led by sonication or other techniques of dispersion. The oxidation of graphene to GO permits efficient separation of layers from each other however to confirm that they could remain in the free state they must be fixed to the surface, separated with surfactant molecules, or covered with the functional groups (Su et al., 2011; Alaferdov et al., 2014). The size of the GO flakes in solution ranges from 10–100 nanometer to 100 μm and for biological applications, scientists utilize those with the least dimension, even 20 nanometers, to assist the entry into the cells (Sanchez et al., 2012). To reinstate the electrical conductivity and other properties of the pristine graphene, it is essential to receive rGO (reduced graphene oxide) which is the product of a decrease of the GO by the usage of reducing agents like chemicals or high temperature. Most of the research's distress GO and its condensed form rGO due to the easiness of preparation, good solubility, and constancy in aqueous solutions in contrast to the other kinds of graphene. Pristine graphene is insoluble in the organic solvents and the monolayers would be inclined to reaggregate in the aqueous solutions (Skoda et al., 2014). The steadiness of the structure of monolayer graphene includes weak interactions of the noncovalent bonding with surfactant molecules. Covalent amendments are accompanied through the creation of novel bonds together with the alteration of sp^2 to sp^3 hybridization. Depending on the kind of amendments, the electrical conductivity of the functionalized graphene is minimized or changed (Park et al., 2009). Several efforts to functionalize graphene employing chemical methods with graphite as an initiating material had permitted so far to get primarily multilayer graphene ($G < 10$) and only in rare cases a single-layer form. Though there are lots of techniques to functionalize graphene, numerous problems yet remain unsolved comprising the quality control, the number of defects, size of the flakes, and the distinction among few-or single-layer graphene. The technique for the mass production of graphene had also not been known (Stankovich, 2007; Xu et al., 2014).

2.4. STRUCTURE DEFECTS

Currently, graphene is one of the most favorable materials in nanotechnology. Though little is known regarding the impact of structural defects of graphene nanostructures due to the synthesis on a huge scale and there is a discrepancy amongst the ideal graphene and its outstanding properties projected by theory and in practice. Numerous laboratory studies presenting its unique properties were carried out on the small, sterilized model samples. As defect-free materials do not exist, it is significant to understand the mechanisms of their creation and the impact on the material properties mainly as when induced on purpose and in an organized manner, they could even become desirable and appropriate for some new solutions (Hernandez et al., 2008; Ricardo et al., 2014). Various kinds of defects could be formed instinctively, could be produced during the production procedure relying on the temperature and chemicals used, and could be artificially introduced to alter the properties of the material. The condensed dimensionality of graphene in these circumstances might be a disadvantage because it raises the number of possible kinds of defects. The probability and because of their creation are not known. Carbon atoms in the 2D honeycomb structure of the graphene have the capability to reorganize and relocate the hexagonal lattice to make non-hexagonal carbon rings. The defects in graphene create the net of carbon=carbon double bonds whose number relies greatly on the technique used for graphene production (Banhart et al., 2011; Rodríguez-Pérez et al., 2013).

Table 2.1. Two Chief Methods with Different Approaches to the Synthesis of Graphene

Method	Description	Advantages/Disadvantages
Top-Down Approaches		
Electrochemical exfoliation	Graphite exfoliation is the sacrificial electrode and gathering graphene from the solution of electrolyte (e.g., H_2SO_4-KOH, surfactant).	Yields the mixture of dissimilar thicknesses of flakes of graphite with the likelihood to segregate few-layer graphene with the help of centrifugation. Surfactant molecules are quite difficult to eradicate and impact the electrochemical and electrical properties of graphene.
Micromechanical cleavage	1st approach utilized to separate graphene from graphite utilizing adhesive tape. Includes repeated cleavage and produces di-, mono-, and the few-layer graphene.	High-quality sheets of graphene. Slow approach, utilized primarily for the investigation of the properties of graphene.
Unzipping CN's (carbon nanotubes)	Few-layer synthesis of graphene via unzipping multi or single-wall CN's utilizing wet chemistry approaches or physical approaches.	The unzipping outcomes in the nano-ribbons of graphene with dissimilar widths are reflected as quasi-1D material with diverse properties as compared to pristine graphene.

Solvent-based exfoliation	Solvent-aided or thermal exfoliation approaches for graphene production from GICs. Exfoliation of the original graphite through sonication in the solvents. Exfoliation of the graphite oxide. (Frequently used approach of the solvent exfoliation and reduction in order to attain graphene comprising Hummers and Offeman's (1958) approach to synthesize the oxide of graphite.)	Costly and dangerous solvents or the surfactant molecules are quite hard to eliminate and affect graphene properties. Increased graphene concentration is escorted by the reduction in the size of flake and upsurge in defect pollution. Exfoliation of the graphite oxide outcomes in rGO with diverse properties as compared to pristine graphene because of high levels of imperfections persuaded by the severe circumstances of the process of production.
Bottom-Up Approaches		
Epitaxial growth	The development of graphene on SiC (silicon carbide) at high temperature and usually in UHV (ultrahigh vacuum) circumstances or diverse atmospheres by the privileged sublimation of Si from silicon carbide surface and the graphitization of C atoms left behind.	Comparatively high quality but seldom ≤ two layers of graphene. Potential transferal of graphene from silicon carbide substrates. Silicon carbide substrates are available commercially but are too costly for commercial uses.
CVD (Chemical vapor deposition)	Development of films of graphene with the help of high-temperature C pyrolysis in the gaseous atmosphere on the substrates of metal. The optimal conditions of the process are dependent on the selected metal substrate. Probable synthesis of the graphene nano-sheets without making the substrate.	Creation of the monolayer graphene but the layers are strongly dependent on several process factors.

Point of symmetry-breaking defects typically take place in the plane of graphene and comprise vacancies and interstitial or substitutional impurities. In any circumstance, the variations in the structure of graphene are because of the lack of one or more than one sp^2 atoms of carbon or the existence of one or more than one diverse atom having sp^3 hybridization. The imperfections can also travel, which has an impact on the properties of the faulty crystal. The imperfections alter the electronic structure and vulnerability to the chemical reactions which change the graphene chemical reactivity (Gao et al., 2011).

A simple example of the in-plane reconstruction is the 5-7-7-5 Stone-Wales flaw which occurs in the revolution of the 2 atoms of carbon by 90° which outcomes in the transformation of 4 6-C rings for two heptagons and two pentagons (Figure 2.5(a)). 2nd simple imperfection related to any material is the lost lattice atom. In order to decrease the system's total energy caused by the non-linear atomic reorganizations (JahnTeller distortion), 2 of the 3 bonds drenches towards the lost atom which causes the creation of 2 novel rings, one ring with the 5 and the second with 9 C atoms (5–9) (Figure 2.5(b)). One more common reorganization of the symmetry of graphene is double vacancies occasioning in one octagon and two pentagons (5-8-5) rather than 4 hexagons which do not disturb the atomic network (Figure 2.5(c)). Except for the 5-8-5 imperfection there are some more probable configurations occasioned by 2 lost atoms; for instance, the revolution of one bond in the octagon creates three heptagons and three pentagons with lower total creation energy (Figure 2.5(d)) or the revolution of the other bond consequential in the 5555-6-7777 imperfection (Figure 2.5(e)). Having more than 2 atoms lost more probable defect configurations can take place (Zsoldos, 2010; Lu et al., 2013).

Other kinds of imperfections alter the graphene charge to a negative or positive or change the complete atomic weight of the entire crystal. Actually, these imperfections damage the mobility of charge but when familiarized deliberately simplify the handling of the transport of charge. These contaminations can be presented into the lattice of carbon by functionalization or substitution of some original atoms of carbon. The graphene doping can be accomplished by intrinsic imperfections modification or with the help of adding distant atoms to the structure of graphene (Araujo et al., 2012; Kuila et al., 2012). Extra atoms substitute one or two atoms of carbon and dependent on the charge or size, considerably larger ones will be moved outward from the plane of graphene. Graphene planarity is decided by the organization of imperfections primarily due to engaging an atom in the in-plane position, which engrosses the 3rd dimension and variations in hybridization. Conversely, when two atoms of carbon traveling over the surface of graphene meet one another and create a dimer, they generally can be integrated into the sp^2 hybridized lattice of carbon but with the cost of local curvature. The graphene properties are dependent on the bonding amongst graphene and foreign atoms. If atomic interactions are governed by the van der Waals forces, the bonding between atoms is quite weak.

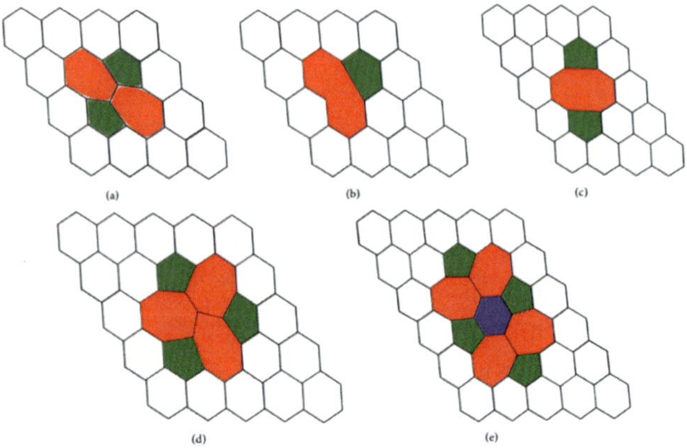

Figure 2.5. Point imperfections of the structure of graphene (a–d). (a) Stone-Wales imperfection SW(5-7-7-5); (b) defect of single vacancy V1(5-9); (c) defect of double vacancy V2(5-8-5); (d) defect of double vacancy V2(555-777); (e) defect of double vacancy V2(5555-6-7777) created from (c).

Source: https://www.hindawi.com/journals/jnm/2014/890246/.

When the foreign atom bonds covalently with the C, the interaction will be quite stronger. Dissimilar strength includes several configurations of bonding which vary the symmetry and takes place primarily on top of the carbon atom, on top of the position of the bridge, or top of the center of the hexagon.

Together with the in-plane changes, the grain-boundary imperfections can also be witnessed, particularly in graphene formed by CVD or mechanical exfoliation. Armchair and zigzag configurations are two chief existent graphene boundary end and exhibit different magnetic and electrical properties. There are also other potential ends, but these two are the most desirable. Synthesis of the graphene structure with distinct edges is not a simple job, and most of the graphene materials are comprised of a mixture of the two motifs. Local variations in the reconstruction kind or persistent elimination of atoms of carbon from the edge offer the imperfections. Armchair edges can normally be converted into the zigzag edges, and all of the intermediates can be well-thought-out imperfect together with the chemical groups that can usually dowse floppy bonds at the edges (Lu et al., 2013). Evidently, dangling bonds and sharp edges can directly disrupt the cell membranes and living organisms irrespective of the properties of graphene.

2.5. THE CLASSIFICATION OF GRAPHENE

Graphene is one layer of atomic carbon and is quite translucent, so the method of characterization is very difficult. The typical methods comprise optical microscope, Raman spectroscopy, TEM, SEM, AFM, etc.

The recognition of sheets of graphene, down to just one layer in the thickness, is probably comprehended via optical microscopy through the contrast of color caused by the interference of light effect on the Silicon oxide (SiO_2) substrate, which is restrained by the layer of graphene. Gao et al. (2011) formed the TCD (total color difference) approach to making it feasible for the classification of large-area samples of graphene. This technique based on an amalgamation of the international commission on illumination color space and reflection spectrum offers reliable and accurate layer identification for quick assessment of the layered variety of graphene by diverse color bands, which provides the likelihood for the physical property measurements and nondestructive recognition of graphene with the help of an optical microscope (Chng et al., 2013).

SEM is also utilized for analysis of graphene. This technique is just like an optical microscope. Normally, it must transfer graphene to the silicon wafer having a particular thickness of SiO_2. The graphene layer can be projected from the depth of color. To recognize films of graphene, Geim's group matched the outcomes of SEM and optical of the large areas on the wafer, struggling to discover the film observable in SEM but not in the optical microscope (Geim, 2011).

Figure 2.6(a) displays the flakes, which is simply identifiable on both optical and SEM images due to the thick region adjoining. The film of graphene gives the more pure contrast in SEM.

AFM is also an efficient way for graphene characterization. Although the graphene thickness is quite thin, it can effortlessly get the morphological characteristic utilizing AFM. From the graphene step on the substrate, it is likely to approximate the number of layers of graphene. Because of the dissimilarities in tip repulsion/attraction amongst insulating substrate and the semi-metallic graphene and under surrounding circumstances by the favored adsorption of the thin layer of water (H_2O) on graphene, it is difficult to attain the theoretical thickness of nearly 0.34 nm (which outcomes in a usual thickness varying from 0.6–1.0 nm for the single layers, as displayed in Figure 2.6(b) and (c)). Thus, the AFM approach for measurement of layer number of the graphene is not precise, and it normally has a quite low throughput.

TEM cannot only notice the morphological characteristic of graphene but can also sum the number of layers of graphene precisely. It is identified that the edges of films of graphene each time fold back, which permits the cross-sectional view of films (as Figure 2.6(d) displays). The reflection of the edges by TEM offers a precise method to tally the number of layers at several positions on the films. Moreover, TEM is frequently aided with the pattern of electron diffraction, which exhibits the hexagonal pattern of the graphene crystal structure (Meyer et al., 2007).

Raman spectroscopy can give an effective and quick way for quality and structure classification of graphene. The observed peaks at 1350 cm^{-1} (given as D band), 1580 cm^{-1} (given as G band), and 2700 cm^{-1} (G' band, also known as a 2D band) are given in the spectrum of samples of graphene. The graphene D band links with the presence of imperfections: the lesser intensity of the D peaks, the lesser imperfections of the layer of graphene. It also links with doping in the graphene. G band is the E_{2g} mode of graphite, which is because of the sp^2-bonded C atoms in the 2D hexagonal layer of graphite. G' band is the double resonance Raman procedure, which comprises the scattering of 2 phonons with contrary momentum about the high-symmetry "K" point in the 1st Brillouin zone of the graphene.

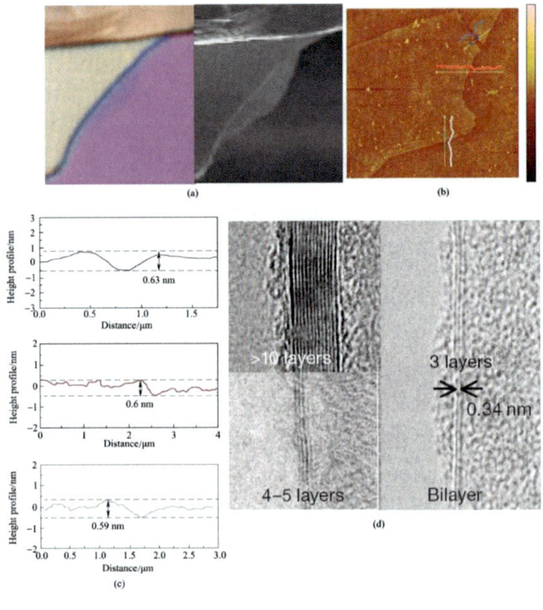

Figure 2.6. (a) Pictures of the thin graphitic flake in scanning electron (right) and optical (left) microscopes. Few-layer graphene is evidently noticeable in SEM but not in the optics; (b) AFM picture of graphene on the substrate of

SiO_2/Si; (c) the graphene height centered on (b); (d) high-magnification TEM picture of graphene.

Source: https://onlinelibrary.wiley.com/doi/abs/10.1002/adma.200904144.

From the Raman spectroscopy, the graphene layers can be deduced. As displayed in Figure 2.7, an additional increase in the layer causes a substantial reduction in the comparative intensity of the G' band. Simultaneously, the G' band will shift to the higher wavenumber and then dowse to that of HOPG, while the intensity ratio $I_{G'}/I_G$ decreases progressively. Therefore, Raman spectroscopy can evidently differentiate the single layer, from the bilayer or few layers. Because of the benefit of nondestructive recognition of layers of graphene and high-throughput, Raman spectroscopy is passionately employed for characterizing graphene currently (Fasolino et al., 2007).

2.6. THE VISION OF GRAPHENE

Research of graphene is emerging at a very quick pace. Since the innovation of graphene, graphene research is unusually extreme. The amount of work on graphene keeps quickly growing over the coming few years. The interest for graphene properties, the objective for its utilization in transparent electrode, sensors, and field-effect transistor (FET), attract scientists to research on graphene with great eagerness. Conversely, it appears that the trouble is not simply being solved currently, and the novel challenges might appear at any instant (Wang et al., 2012).

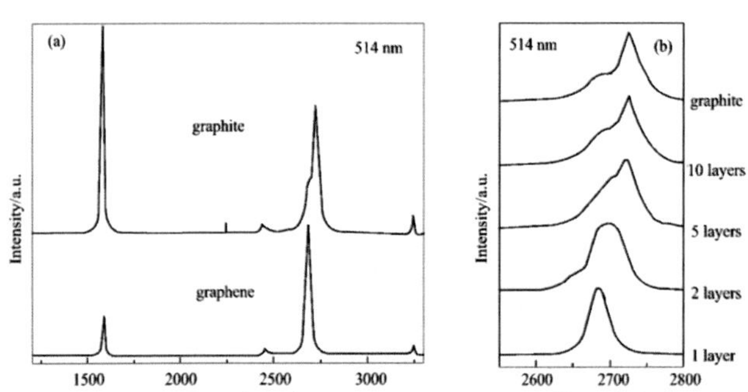

Figure 2.7. (a) The contrast of spectra of Raman at 514 nanometers for bulk graphene and graphite; (b) advancement of a 2D band at 514 nanometers with the number of layers.

Source: https://pubmed.ncbi.nlm.nih.gov/19350030/.

First, a new method for high quality, large area, fast growth, and low temperature shall be discovered in the future. Even though numerous approaches were established for the synthesis of graphene, quality, throughput, and scale still need further enhancement (Hummers and Offeman, 1958; Liao et al., 2011).

Second, the majority of the current characterization approaches must transfer graphene to the quantified substrate, which reduces the characterization efficiency, like silicon with a particular thickness of the intrinsic oxide. Although Raman spectroscopy offers a fast recognition of layers of graphene, once the film of graphene is up to various layers, it is problematic to differentiate it from graphite. An effective and quick identification approach must be discovered (Chatterjee et al., 2014).

Lastly, thanks to the current fast development of the transfer method, graphene can be accurately and easily transferred to the other substrates. Conversely, how to incorporate graphene into current-day electronic devices is on the way, which is crucial for the application of graphene. Graphene-based devices must be passionately studied further (Yue et al., 2012).

Even with the presence of several difficulties, it is supposed that the appropriate synthetic methods can be discovered soon. The quality and size of graphene are improving significantly in current years, and there is not any hesitation that it will be improved further in the future. With the efforts and in-depth study by scholars, graphene, and the composite materials of graphene will be made practical for the national economy (Sasidharan et al., 2011, 2012).

2.7. GRAPHENE AS A BIOMATERIAL

Biomaterial are substances that, as a contrast to a drug or an amalgamation of synthetic or natural substances, are envisioned to interface with the biological systems in order to assess, treat, enhance, or substitute any tissue, function, or organ of the body. Tissue engineering is an amalgamation of the information regarding the cellular environment, cells, and materials utilized to regenerate or enhance tissues and their functionality. No matter what kind of biomaterial has been used, appropriate support for tissue engineering must meet specific requirements. Firstly, any biomaterial should be biocompatible (Chen et al., 2012; Zhou et al., 2012). After embedding the scaffold cannot tempt immune response which might decrease the process of healing and permit the cells to obey, multiply, migrate, and then function normally. As the implants are not meant to be everlasting, the biomaterial must be biodegradable and permit the cells to yield their own ECM (extracellular matrix). All of the byproducts left behind after the degradation must not be poisonous and must leave the body without causing any damage. Ideal biomaterial must have quantified mechanical properties with porous architecture in order to permit cell infiltration and sufficient diffusion of the nutrients. And ultimately the manufacturing method should be cheap and feasible for the scale-up processing and must be made accessible to the clinicians (O'brien, 2011; Li et al., 2012). Selecting the right natural or synthetic biomaterial generally plays a crucial part in tissue engineering. The properties of material like surface chemistry and topography, porosity, or roughness can modify the response of the cell and it is quite that biomaterial surface usually plays a significant role in manipulating cell phenotype and the other growth crucial factors (Jastrzębska et al., 2012; Hu and Zhou, 2013). The proliferation, cell viability, and fate are strongly dependent on the ECM. A platform that mimics the ECM and governs the behavior of cells and the progression of tissue would be crucial in clinical applications. Because of the superlative properties, graphene has discovered the potential in a broad range of areas, comprising tissue engineering. Graphene can also offer an ideal surface for the cell culture due to its good biocompatibility, high elasticity, chemical inertness, electrical conductivity, and flexibility. Materials based on graphene are favorable tools for tissue engineering, primarily due to its condensed, regular structure vulnerable to modification and functionalization, and aptitude to bind a range of molecules like proteins, drugs, or DNA (Pinto et al., 2013). Together, with the point that C is the fundamental element of the biological structures, nanomaterials based on graphene can act as the bioactive scaffold and structural strengthening for other biomaterials presently being utilized in tissue engineering (Zhang et al., 2012; Li et al., 2014).

Graphene and the derivatives of graphene have been employed as the substrate for stem cells in regenerative medicine. To utilize stem cells in tissue engineering and transplantation, it is essential to stimulate them appropriately and provide appropriate natural or synthetic growth conditions. An extra advantage would be the scaffold material which generally has a governable and collaborating interface with the cells. There are primarily three kinds of stem cells utilized in the regenerative medicine: pluripotent ESC (embryonic stem cells) separated from the inner cell mass of blastocyst ESC, ASC (adult stem cells) with comparable regenerative potential but decreased capability to distinguish, and iPS (induced pluripotent stem cells) produced in vitro from the somatic cells (Martino et al., 2012; Feng et al., 2013). The Mesenchymal stem cells can normally be separated from numerous adult tissues and can distinguish into adipocytes, chondrocytes, or osteocytes dependent on the conditions of growth. The first-ever research of Kalbacova and his coworkers exhibited that the graphene produced through CVD on the foils of copper is not poisonous for human osteoblasts or hMSC (mesenchymal stem cells) but is even appropriate for their proliferation, growth, and later distinction into osteoblasts (Kalbacova et al., 2010). Research of Nayak et al. also displayed that graphene does not affect the viability, morphology, or growth of the human mesenchymal stem cells. Furthermore, they also proved that MSC distinguishes into the bone cells in a similar way to samples with the accumulation of suitable growth factors applied for osteogenic differentiation (Nayak et al., 2011; Bianco, 2013). A bit later Lee et al. recommended the theory because of which human mesenchymal stem cells are expected to distinguish into osteoblasts on the graphene and derivatives of graphene by linking the strong non-covalent graphene obligatory capabilities towards diverse growth agents and the extent of π-πstacking (Lee et al., 2011; Wang et al., 2011).

The biocompatibility of graphene to mouse iPS cells with good observance and proliferation was proved by Chen and coworkers together with the inference to the great prospective of graphene covered materials as the platforms for several medical applications and not just for hard tissue engineering (Chen et al., 2012). For brain repair and neuronal regeneration, it is dangerous to tempt the proper hNSC discrepancy towards neurons. The study revealed that the substrates of graphene have excellent electrochemical and electrical properties and aid the development of functional neurons and enhance their electrical signaling and performance (Li et al., 2013; Yang et al., 2013). The growth of cells can be encouraged by the variation of surface charge via the graphene contact with voltage-gated ion channels. During research, Park and his coworkers noticed the encouraging adhesion of cell, attachment, and improved neural discrepancy on the films of graphene contrary to the conventional substrates like glass (Park et al., 2011; Shen et al., 2012). Several reports recommend that graphene is a brilliant material for the research with adherent cells, particularly in the film form where it can display good biocompatibility without any viability inhibition. The study on adherent fibroblasts cells NIH-3T3 developed on the films of graphene and the human cancer epithelial cells A549 outcomes in good attachment and promising growth with no inducing harmful effects whereas improving cellular functions (Ryoo et al., 2010; Chang et al., 2011). The research also demonstrates that the GO paper improves the proliferation and attachment of mammalian cells with no tempting cytotoxic effects (Ruiz et al., 2011).

REFERENCES

1. Alaferdov, A. V., Gholamipour-Shirazi, A., Canesqui, M. A., Danilov, Y. A., & Moshkalev, S. A., (2014). Size-controlled synthesis of graphite nanoflakes and multi-layer graphene by liquid-phase exfoliation of natural graphite. *Carbon, 69*, 525–535.
2. Allen, M. J., Tung, V. C., & Kaner, R. B., (2010). Honeycomb carbon: A review of graphene. *Chemical Reviews, 110*(1), 132–145.
3. Araujo, P. T., Terrones, M., & Dresselhaus, M. S., (2012). Defects and impurities in graphene-like materials. *Materials Today, 15*(3), 98–109.
4. Bae, S., Kim, H., Lee, Y., Xu, X., Park, J. S., Zheng, Y., & Kim, Y. J., (2010). Roll-to-roll production of 30-inch graphene films for transparent electrodes. *Nature Nanotechnology, 5*(8), 574.
5. Banhart, F., Kotakoski, J., & Krasheninnikov, A. V., (2011). Structural defects in graphene. *ACS Nano, 5*(1), 26–41.
6. Bianco, A., (2013). Graphene: Safe or toxic? The two faces of the medal. *Angewandte Chemie International Edition, 52*(19), 4986–4997.
7. Bitounis, D., Ali-Boucetta, H., Hong, B. H., Min, D. H., & Kostarelos, K., (2013). Prospects and challenges of graphene in biomedical applications. *Advanced Materials, 25*(16), 2258–2268.
8. Boehm, H. P., Setton, R., & Stumpp, E., (1994). Nomenclature and terminology of graphite intercalation compounds (IUPAC Recommendations 1994). *Pure and Applied Chemistry, 66*(9), 1893–1901.
9. Chang, Y., Yang, S. T., Liu, J. H., Dong, E., Wang, Y., Cao, A., & Wang, H., (2011). *In Vitro* toxicity evaluation of graphene oxide on A549 cells. *Toxicology Letters, 200*(3), 201–210.
10. Chatterjee, N., Eom, H. J., & Choi, J., (2014). A systems toxicology approach to the surface functionality control of graphene-cell interactions. *Biomaterials, 35*(4), 1109–1127.
11. Chen, G. Y., Pang, D. P., Hwang, S. M., Tuan, H. Y., & Hu, Y. C., (2012). A graphene-based platform for induced pluripotent stem cells culture and differentiation. *Biomaterials, 33*(2), 418–427.
12. Chen, G. Y., Yang, H. J., Lu, C. H., Chao, Y. C., Hwang, S. M., Chen, C. L., & Hu, Y. C., (2012). Simultaneous induction of autophagy and toll-like receptor signaling pathways by graphene oxide. *Biomaterials, 33*(27), 6559–6569.

13. Chia, J. S. Y., Tan, M. T., SimKhiew, P., Chin, J. K., Lee, H., Bien, D. C. S., & Siong, C. W., (2013). Facile synthesis of few-layer graphene by mild solvent thermal exfoliation of highly oriented pyrolytic graphite. *Chemical Engineering Journal, 231*, 1–11.
14. Chng, E. L. K., & Pumera, M., (2013). The toxicity of graphene oxides: Dependence on the oxidative methods used. *Chemistry: A European Journal, 19*(25), 8227–8235.
15. Edwards, R. S., & Coleman, K. S., (2013). Graphene synthesis: relationship to applications. *Nanoscale,5*(1), 38–51.
16. Eigler, S., & Hirsch, A., (2014). Chemistry with graphene and graphene oxide—challenges for synthetic chemists. *Angewandte Chemie International Edition, 53*(30), 7720–7738.
17. Eremina, E. A., Kaplin, A. V., Eliseev, A. A., Sidorov, A. V., Radzhabzoda, S. S., Grigor'eva, A. V., & Gudilin, E. A., (2018). Multifunctional composites based on graphite oxide, doxorubicin, and magnetic nanoparticles for targeted drug delivery. *Nanotechnologies in Russia, 13*(3, 4), 152–160.
18. Falcao, E. H., & Wudl, F., (2007). Carbon allotropes: Beyond graphite and diamond. *Journal of Chemical Technology and Biotechnology: International Research in Process, Environmental and Clean Technology, 82*(6), 524–531.
19. Fasolino, A., Los, J. H., & Katsnelson, M. I., (2007). Intrinsic ripples in graphene. *Nature Materials, 6*(11), 858–861.
20. Feng, L., Wu, L., & Qu, X., (2013). New horizons for diagnostics and therapeutic applications of graphene and graphene oxide. *Advanced Materials, 25*(2), 168–186.
21. Ferrari, A. C., Meyer, J. C., Scardaci, V., Casiraghi, C., Lazzeri, M., Mauri, F., & Geim, A. K., (2006). Raman spectrum of graphene and graphene layers. *Physical Review Letters, 97*(18), 187401.
22. Fuhrer, M. S., Lau, C. N., & MacDonald, A. H., (2010). Graphene: Materially better carbon. *MRS Bulletin, 35*(4), 289–295.
23. Gao, X., Wang, Y., Liu, X., Chan, T. L., Irle, S., Zhao, Y., & Zhang, S. B., (2011). Regioselectivity control of graphene functionalization by ripples. *Physical Chemistry Chemical Physics, 13*(43), 19449–19453.
24. Geim, A. K., (2009). Graphene: Status and prospects. *Science, 324*(5934), 1530–1534.

25. Geim, A. K., (2011). Nobel Lecture: Random walk to graphene. *Reviews of Modern Physics, 83*(3), 851.
26. Hernandez, Y., Nicolosi, V., Lotya, M., Blighe, F. M., Sun, Z., De, S., & Boland, J. J., (2008). High-yield production of graphene by liquid-phase exfoliation of graphite. *Nature Nanotechnology, 3*(9), 563–568.
27. Hu, X., & Zhou, Q., (2013). Health and ecosystem risks of graphene. *Chemical Reviews, 113*(5), 3815–3835.
28. Hummers, Jr. W. S., & Offeman, R. E., (1958). Preparation of graphitic oxide. *Journal of the American Chemical Society, 80*(6), 1339–1339.
29. Iijima, S., (1991). Helical microtubules of graphitic carbon. *Nature, 354*(6348), 56–58.
30. Jastrzębska, A. M., Kurtycz, P., & Olszyna, A. R., (2012). Recent advances in graphene family materials toxicity investigations. *Journal of Nanoparticle Research, 14*(12), 1320.
31. Kalbacova, M., Broz, A., Kong, J., & Kalbac, M., (2010). Graphene substrates promote adherence of human osteoblasts and mesenchymal stromal cells. *Carbon, 48*(15), 4323–4329.
32. Kuila, T., Bose, S., Mishra, A. K., Khanra, P., Kim, N. H., & Lee, J. H., (2012). Chemical functionalization of graphene and its applications. *Progress in Materials Science, 57*(7), 1061–1105.
33. Lee, W. C., Lim, C. H. Y., Shi, H., Tang, L. A., Wang, Y., Lim, C. T., & Loh, K. P., (2011). Origin of enhanced stem cell growth and differentiation on graphene and graphene oxide. *ACS Nano, 5*(9), 7334–7341.
34. Li, N., Cheng, Y., Song, Q., Jiang, Z., Tang, M., & Cheng, G., (2014). Graphene meets biology. *Chinese Science Bulletin, 59*(13), 1341–1354.
35. Li, N., Zhang, Q., Gao, S., Song, Q., Huang, R., Wang, L., & Cheng, G., (2013). Three-dimensional graphene foam as a biocompatible and conductive scaffold for neural stem cells. *Scientific Reports, 3*(1), 1–6.
36. Li, X., Cai, W., An, J., Kim, S., Nah, J., Yang, D., & Banerjee, S. K., (2009). Large-area synthesis of high-quality and uniform graphene films on copper foils. *Science, 324*(5932), 1312–1314.
37. Li, Y., Liu, Y., Fu, Y., Wei, T., Le Guyader, L., Gao, G., & Chen, C., (2012). The triggering of apoptosis in macrophages by pristine graphene through the MAPK and TGF-beta signaling pathways. *Biomaterials, 33*(2), 402–411.

38. Liao, K. H., Lin, Y. S., Macosko, C. W., & Haynes, C. L., (2011). Cytotoxicity of graphene oxide and graphene in human erythrocytes and skin fibroblasts. *ACS Applied Materials and Interfaces, 3*(7), 2607–2615.
39. Lu, J., Bao, Y., Su, C. L., & Loh, K. P., (2013). Properties of strained structures and topological defects in graphene. *ACS Nano, 7*(10), 8350–8357.
40. Mao, H. Y., Laurent, S., Chen, W., Akhavan, O., Imani, M., Ashkarran, A. A., & Mahmoudi, M., (2013). Graphene: Promises, facts, opportunities, and challenges in nanomedicine. *Chemical Reviews, 113*(5), 3407–3424.
41. Martino, S., D'Angelo, F., Armentano, I., Kenny, J. M., & Orlacchio, A., (2012). Stem cell-biomaterial interactions for regenerative medicine. *Biotechnology Advances, 30*(1), 338–351.
42. Meca, E., Lowengrub, J., Kim, H., Mattevi, C., & Shenoy, V. B., (2013). Epitaxial graphene growth and shape dynamics on copper: Phase-field modeling and experiments. *Nano Letters, 13*(11), 5692–5697.
43. Meyer, J. C., Geim, A. K., Katsnelson, M. I., Novoselov, K. S., Booth, T. J., & Roth, S., (2007). The structure of suspended graphene sheets. *Nature, 446*(7131), 60–63.
44. Mohanty, N., Moore, D., Xu, Z., Sreeprasad, T. S., Nagaraja, A., Rodriguez, A. A., & Berry, V., (2012). Nanotomy-based production of transferable and dispersible graphene nanostructures of controlled shape and size. *Nature Communications, 3*(1), 1–8.
45. Nayak, T. R., Andersen, H., Makam, V. S., Khaw, C., Bae, S., Xu, X., & Ozyilmaz, B., (2011). Graphene for controlled and accelerated osteogenic differentiation of human mesenchymal stem cells. *ACS Nano, 5*(6), 4670–4678.
46. Novoselov, K. S., (2011). Nobel lecture: Graphene: Materials in the flatland. *Reviews of Modern Physics, 83*(3), 837.
47. Novoselov, K. S., Geim, A. K., Morozov, S. V., Jiang, D., Zhang, Y., Dubonos, S. V., & Firsov, A. A., (2004). Electric field effect in atomically thin carbon films. *Science, 306*(5696), 666–669.
48. O'brien, F. J., (2011). Biomaterials and scaffolds for tissue engineering. *Materials Today, 14*(3), 88–95.

49. Obraztsov, A. N., Obraztsova, E. A., Tyurnina, A. V., & Zolotukhin, A. A., (2007). Chemical vapor deposition of thin graphite films of nanometer thickness. *Carbon, 45*(10), 2017–2021.
50. Obst, A., Šingliar, U., Pätzold, C., Müller, A., & Bertau, M., (2017). Synthesis and characterization of in situ doped silicon nanoparticles. *Silicon, 9*(2), 135–145.
51. Park, S. Y., Park, J., Sim, S. H., Sung, M. G., Kim, K. S., Hong, B. H., & Hong, S., (2011). Enhanced differentiation of human neural stem cells into neurons on graphene. *Advanced Materials, 23*(36), H263–H267.
52. Park, S., & Ruoff, R. S., (2009). Chemical methods for the production of graphenes. *Nature Nanotechnology, 4*(4), 217–224.
53. Park, S., An, J., Jung, I., Piner, R. D., An, S. J., Li, X., & Ruoff, R. S., (2009). Colloidal suspensions of highly reduced graphene oxide in a wide variety of organic solvents. *Nano Letters, 9*(4), 1593–1597.
54. Pinto, A. M., Goncalves, I. C., & Magalhaes, F. D., (2013). Graphene-based materials biocompatibility: A review. *Colloids and Surfaces B: Biointerfaces, 111*, 188–202.
55. Ricardo, K. B., Sendecki, A., & Liu, H., (2014). Surfactant-free exfoliation of graphite in aqueous solutions. *Chemical Communications, 50*(21), 2751–2754.
56. Rodríguez-Pérez, L., Herranz, M. Á., & Martín, N., (2013). The chemistry of pristine graphene. *Chemical Communications, 49*(36), 3721–3735.
57. Ruan, G., Sun, Z., Peng, Z., & Tour, J. M., (2011). Growth of graphene from food, insects, and waste. *ACS Nano, 5*(9), 7601–7607.
58. Ruiz, O. N., Fernando, K. S., Wang, B., Brown, N. A., Luo, P. G., McNamara, N. D., & Bunker, C. E., (2011). Graphene oxide: A nonspecific enhancer of cellular growth. *ACS Nano, 5*(10), 8100–8107.
59. Ryoo, S. R., Kim, Y. K., Kim, M. H., & Min, D. H., (2010). Behaviors of NIH-3T3 fibroblasts on graphene/carbon nanotubes: proliferation, focal adhesion, and gene transfection studies. *ACS Nano, 4*(11), 6587–6598.
60. Sanchez, V. C., Jachak, A., Hurt, R. H., & Kane, A. B., (2012). Biological interactions of graphene-family nanomaterials: An interdisciplinary review. *Chemical Research in Toxicology, 25*(1), 15–34.

61. Sasidharan, A., Panchakarla, L. S., Chandran, P., Menon, D., Nair, S., Rao, C. N. R., & Koyakutty, M., (2011). Differential nano-bio interactions and toxicity effects of pristine versus functionalized graphene. *Nanoscale, 3*(6), 2461–2464.

62. Sasidharan, A., Panchakarla, L. S., Sadanandan, A. R., Ashokan, A., Chandran, P., Girish, C. M., & Koyakutty, M., (2012). Hemocompatibility and macrophage response of pristine and functionalized graphene. *Small, 8*(8), 1251–1263.

63. Schedin, F., Geim, A. K., Morozov, S. V., Hill, E. W., Blake, P., Katsnelson, M. I., & Novoselov, K. S., (2007). Detection of individual gas molecules adsorbed on graphene. *Nature Materials, 6*(9), 652–655.

64. Segal, M., (2012). Material history: Learning from silicon. *Nature, 483*(7389), S43–S44.

65. Shen, H., Zhang, L., Liu, M., & Zhang, Z., (2012). Biomedical applications of graphene. *Theranostics, 2*, 283–294.

66. Singh, V., Joung, D., Zhai, L., Das, S., Khondaker, S. I., & Seal, S., (2011). Graphene based materials: Past, present, and future. *Progress in Materials Science, 56*(8), 1178–1271.

67. Skoda, M., Dudek, I., Jarosz, A., & Szukiewicz, D., (2014). Graphene: One material, many possibilities—application difficulties in biological systems. *Journal of Nanomaterials, 2014.*

68. Smalley, R. E., (1997). Discovering the fullerenes (Nobel lecture). *Angewandte Chemie International Edition in English, 36*(15), 1594–1601.

69. Soldano, C., Mahmood, A., & Dujardin, E., (2010). Production, properties, and potential of graphene. *Carbon, 48*(8), 2127–2150.

70. Stankovich, S., Dikin, D. A., Piner, R. D., Kohlhaas, K. A., Kleinhammes, A., Jia, Y., & Ruoff, R. S., (2007). Synthesis of graphene-based nanosheets via chemical reduction of exfoliated graphite oxide. *Carbon, 45*(7), 1558–1565.

71. Su, C. Y., Lu, A. Y., Xu, Y., Chen, F. R., Khlobystov, A. N., & Li, L. J., (2011). High-quality thin graphene films from fast electrochemical exfoliation. *ACS Nano, 5*(3), 2332–2339.

72. Sun, Z., Yan, Z., Yao, J., Beitler, E., Zhu, Y., & Tour, J. M., (2010). Growth of graphene from solid carbon sources. *Nature, 468*(7323), 549–552.

73. Van, N. R., (2012). Production: Beyond sticky tape. *Nature, 483*(7389), S32–S33.
74. Wang, D., Wang, L., Dong, X., Shi, Z., & Jin, J., (2012). Chemically tailoring graphene oxides into fluorescent nanosheets for Fe^{3+} ion detection. *Carbon, 50*(6), 2147–2154.
75. Wang, Y., Li, Z., Wang, J., Li, J., & Lin, Y., (2011). Graphene and graphene oxide: Biofunctionalization and applications in biotechnology. *Trends in Biotechnology, 29*(5), 205–212.
76. Wei, D., & Liu, Y., (2010). Controllable synthesis of graphene and its applications. *Advanced Materials, 22*(30), 3225–3241.
77. Wei, D., Liu, Y., Cao, L., Fu, L., Li, X., Wang, Y., & Yu, G., (2007). A magnetism-assisted chemical vapor deposition method to produce branched or iron-encapsulated carbon nanotubes. *Journal of the American Chemical Society, 129*(23), 7364–7368.
78. Xu, J., Dang, D. K., Liu, X., Chung, J. S., Hur, S. H., Choi, W. M., & Kohl, P. A., (2014). Liquid-phase exfoliation of graphene in organic solvents with addition of naphthalene. *Journal of Colloid and Interface Science, 418*, 37–42.
79. Yan, L., Zhao, F., Li, S., Hu, Z., & Zhao, Y., (2011). Low-toxic and safe nanomaterials by surface-chemical design, carbon nanotubes, fullerenes, metallofullerenes, and graphenes. *Nanoscale,3*(2), 362–382.
80. Yang, M., Yao, J., & Duan, Y., (2013). Graphene and its derivatives for cell biotechnology. *Analyst,138*(1), 72–86.
81. Yang, W., Chen, G., Shi, Z., Liu, C. C., Zhang, L., Xie, G., & Watanabe, K., (2013). Epitaxial growth of single-domain graphene on hexagonal boron nitride. *Nature Materials, 12*(9), 792–797.
82. Yue, H., Wei, W., Yue, Z., Wang, B., Luo, N., Gao, Y., & Su, Z., (2012). The role of the lateral dimension of graphene oxide in the regulation of cellular responses. *Biomaterials, 33*(16), 4013–4021.
83. Zhang, H., Gruener, G., & Zhao, Y., (2013). Recent advancements of graphene in biomedicine. *Journal of Materials Chemistry B,1*(20), 2542–2567.
84. Zhang, Y., Nayak, T. R., Hong, H., & Cai, W., (2012). Graphene: A versatile nanoplatform for biomedical applications. *Nanoscale, 4*(13), 3833–3842.

85. Zhou, H., Zhao, K., Li, W., Yang, N., Liu, Y., Chen, C., & Wei, T., (2012). The interactions between pristine graphene and macrophages and the production of cytokines/chemokines via TLR-and NF-κB-related signaling pathways. *Biomaterials, 33*(29), 6933–6942.
86. Zhu, Y., Murali, S., Cai, W., Li, X., Suk, J. W., Potts, J. R., & Ruoff, R. S., (2010). Graphene and graphene oxide: Synthesis, properties, and applications. *Advanced Materials, 22*(35), 3906–3924.
87. Zsoldos, I., (2010). Effect of topological defects on graphene geometry and stability. *Nanotechnology, Science and Applications, 3*, 101.

CHAPTER 3

Applications of Graphene-Based Materials in Electronic Devices

CONTENTS

3.1. INTRODUCTION

In 1947, despite the band structure of graphene, an ideal two-dimensional (2D) sheet accompanying carbon atoms organized in a honeycomb crystal lattice had been anticipated, such a thin one-atomic-layer material has been really cumbersome to obtain, and monolayer graphene was normally expected to be unstable in nature (Wallace, 1947; Neto et al., 2009). In 1991, carbon nanotube (CNT), which is a one-dimensional (1D), carbon material was found (Iijima, 1991). The semiconducting or metallic electronic band structure is manifested by CNT relying upon the chirality of the tube, also shares enough basic characteristics and functions with graphene. The chirality of CNTs can be inferred and concluded from various orientations of a rolled graphene sheet (Dresselhaus et al., 1998). At last, in 2004, the physical exfoliation method well synthesized a sole layer of graphene that is thermodynamically stable, and the verification of the Dirac band structure was done by the demonstration of an ambipolar transport (Novoselov et al., 2004). The chirality-dependent semiconducting and metallic properties analogous to CNT was also verified by the successive experiments on graphene nanoribbons (GNRs). Over the last decade, therefore, graphene has become the dominant material in the fields of chemistry, physics, engineering, and bioscience, and it has drawn the attention of researchers (Zhang et al., 2005; Du et al., 2008).

The carbon atoms in graphene constitute out-of-plane π-bonds and in-plane sp^2-hybridized σ-bonds at every site. Robust lattice graphene is provided by the formers consequently in high-stress tolerance, while graphene having a unique electronic band structure with a linear dispersion at the Dirac points situated at the edge of the first Brillouin zone (i.e., K points in the reciprocal lattice of graphene) is presented by the recent. In favor of the Dirac dispersion relation, many impressive characteristics of graphene have been presented and illustrated. The Dirac fermions have the capability to move with high speed ($\sim 10^6$ m s^{-1}) and are massless. The practically recorded values of mobility for suspended graphene have exceeded 200,000 cm^2 V^{-1} s^{-1}, and for chemical vapor deposited (CVD) grown (on a born-nitride substrate), graphene has exceeded 35,000 cm^2 V^{-1} s^{-1}. It has been recorded recently that for hexagonal-boron-nitride (hBN) heterostructure with graphene, electron mobility is 500,000 cm^2 V^{-1} s^{-1} at low room temperature, and up to 100,000 cm^2 V^{-1} s^{-1} at room temperature (*T*). Moreover, other significant features have been observed experimentally and theoretically, such as the anomalous integer quantum Hall effect, quantum electrodynamics effects, electro-optical effects, Klein paradox, and extremely high thermal conductivity. Thus, the focus of this chapter will be on graphene's possible applications in electronic devices like RF transistors chip-interconnects, and others based on electro-optical effects (Bolotin et al., 2008; Gannett et al., 2011).

In spite of that, because there is a possibility of the fabrication of GNRs from bottom-up or top-down approaches to make a bandgap for the graphene-based material by quantum confinement effect and the possibility of few uncommon behaviors with spin polarizations under the electric field, and external magnetic field (Katsnelson et al., 2006; Cheianov et al., 2007). Therefore, we sum up some research studies on tunneling field-effect transistors (TFETs), metal-oxide-semiconductor field-effect transistors (MOSFETs), spintronic devices based on GNRs, and resonant-tunneling diodes (RTDs).

3.2. GRAPHENE AND GNRS

Since experimental measurements to date show strong sample-dependent performances, the most important prospect of graphene devices is the fabrication issues from the application point of view. The highest quality of graphene that has been produced still depends on the mechanical exfoliation of graphite (Novoselov et al., 2005; Mayorov et al., 2011). This strategy delivers best for experimental and laboratory purposes which asked for the reasonably large (micron size) and highest quality graphene flakes. Mechanically exfoliated graphene, being deposited or suspended on an appropriate substrate such as boron nitride, mostly exhibits the high-quality transport measurements along with the highest mobility. There are also alternate ways to manufacture reasonably high-quality graphene by unzipping single-walled carbon nanotubes (SWCNTs). Suspensions can be achieved by this method with well-defined dissemination of graphene platelets and also this method gives control over chemical functionalization plus the good quality of the edges (Jiao et al., 2009; Dean et al., 2010). Chemical exfoliations of graphite with organic solvents have also been used and can create high-quality samples at even low-cost than unzipping SWCNTs. Bae et al. (2010) have fabricated exceptionally larger graphene films of square meter size with the help of a so-called roll-to-roll strategy in collusion with Samsung. The films have been shifted onto 200 mm Si wafers good enough for direct industrial production. They can already be consumed in transparent conductive coating applications, e.g., touch screen devices, regardless of some additions of thicker layers and more flaws found on these large films. Novoselov et al. (2005) recently discussed a review on numerous fabrication methods and their figure of merits.

As given in the introduction section, the latest cited mobility of graphene is over 200,000 $cm^2 V^{-1} s^{-1}$, which is near to the theoretical predictions. The large-area graphene samples fabricated with mechanical exfoliations measured this ultra-high mobility of Dirac particles resulting in no bandgap. However, a bandgap is essential for practical devices in order to have an off-state (Hernandez et al., 2008; Kosynkin et al., 2009). Furthermore, practically often a high-quality GNR is more preferable instead of a large-area graphene flake, by reason of they can be used as p-n junction devices or transistors. Transverse confinements can easily introduce the desired bandgap on GNRs. Zigzag graphene nanoribbons (ZGNRs) and armchair graphene nanoribbons (AGNRs) both are predicted by the tight-binding model of graphene to have a bandgap that is nearly inversely proportional to the width of the nanoribbon (Han et al., 2007; Novoselov and Fal, 2012). AGNRs are theoretically semiconducting and their bandgap change with the width in three disparate families. Whereas, ZGNRs are always theoretically metallic with ferromagnetic edge states directing to differing spins. Nevertheless, these preliminary theoretical models exclude chaos which may revise the characteristics relatively (Table 3.1) (Son et al., 2006; Balandin et al., 2008).

Table 3.1. Summary of GNR Fabrication Techniques

GNR Fabrication Techniques	**Main Characteristics**
Mechanical exfoliation of graphite	High mobility, small crystalline flakes (several microns), appropriate for lab objectives
Thermal epitaxial growths from SiC	Much lower mobility because of smaller crystal sizes and more defects. It may be favorable for applications where SiC semiconductor substrates are needed
Chemical exfoliation of graphite	Lower mobility because of smaller crystalline flakes (few microns) enough for labs, and introduction of the chemical complex. Comprehensive sample size can be much greater than mechanical exfoliation techniques
Roll to roll technique with chemical vapor deposition	Low mobility and higher defects concentrations. Applicable for large-scale industrial fabrications of transparent conducting surfaces

In nanoribbons, the opening of a bandgap has been confirmed experimentally for widths down to nearly 1 nm, and band gaps more than 0.2 eV are shown for widths <20 nm. Conventional GNRs have defected edges and their widths change along their lengths. Often, these disorders smear out some disparity in the bandgap of ZGNRs and AGNRs. For practical devices, larger band gaps in GNRs are more preferable; however, this simultaneously makes the valence resulting in the more parabolic conduction bands (CBs) around the Dirac point and raises the effective mass of the charge carriers. Consequently, lessening the mobility of the charge carrier in graphene (Huang et al., 2007; Kim and Kim, 2008).

The conclusion derived from the conventional semiconductor theory is that mobility decreases when the bandgap increases, and the same as for GNRs. This indicates that as compared to silicon devices (1.1 eV), the mobility of electrons in GNRs with a bandgap is supposed to be reduced than bulk silicon and not inevitably higher than silicon channels in traditional MOSFETs (Areshkin et al., 2007; Li et al., 2008). The truly existing mobilities determined in different experiments exhibit approximate 1500 $cm^2 V^{-1} s^{-1}$ for nanoribbons with 14 nm width, and<200 $cm^2 V^{-1} s^{-1}$ for nanoribbon with widths between 1 and 10 nm, and these chiefly coming from defects and edge disorders (Son et al., 2006; Fiori and Iannaccone, 2007). Hence, to overcome defects and disorder during the fabrication process, better fabrication technologies are needed in order for graphene to be more remarkable to the silicon counterparts. In GNRs, capacity for smaller bandgap also lessens the contemporary on-off ratios demanded the efficient operation of FETs. Regardless of encouraging development in graphene transistors, still, some ingenious problems are faced by most of the graphene transistors, i.e., peculiar current-saturation behaviors at high frequency and high off-state currents. It is inviting to make narrower GNRs in order to make better the off-state current but this makes degenerate the mobility even further and edge disorders become superior (Wakabayashi, 2001; Liang et al., 2007; Wang et al., 2008).

3.3. GRAPHENE DEVICES

3.3.1. Electro-Optic Devices

Even though in the former high mobility of electrons, the deficiency of bandgap in graphene has limited its prospects for taken over silicon in MOSFETs. A MOSFET cannot be switched off without a bandgap which in turn a poor current switching ratio (on current to off current, i.e., I_{on}/I_{off}). Even though a switching ratio >10^6 has been exhibited by the GNR FETs with sub-10 nm width, the mobility has been badly jeopardized. In addition, reliable fabrication of such narrow GNRs is a challenge. To induce the bandgap, restricted suitability for graphene FETs (GFETs) has also been indicated by other schemes, e.g., bilayer graphene and strained GNR. Therefore, to accomplish required switching, traditional phenomenon such as pseudospins and negative refraction have been recommended (Guo et al., 2008; Teong et al., 2009; Jiao et al., 2010).

Negative refraction is exhibited by a p-n junction in graphene considering an optical analog of the Veselago lens, whose focus is on the obliquely incident electron waves. For monolayer (bilayer) graphene, normally incident wave utterly transmits (reflects) via the junction, a phenomenon called Klein tunneling. Thus, graphene permits devices with a unique harnessed over electron waves, for example, in photonic systems, electromagnetic (EM) waves are controlled (Guo et al., 2008; Zeng et al., 2011a, b). However, electrons have smaller de Broglie wavelength as compared to photons, a characteristic that gives higher resolution in an electron microscope, and graphene electro-optic devices can be extremely less than their optoelectronic counterparts because device size is restricted by the operating wavelength (Obradovic et al., 2006; Fang et al., 2008). An optimized solution for high-speed low-power electronic components is offered by the high mobility of electrons achieved with the electro-optic effect and the mean-free path (mfp) longer than the aimed channel lengths of eventual computing devices. The following three types of device investigations have yet been engendered by the electro-optic effect in graphene (Table 3.2) (Zhao et al., 2009; Chin et al., 2010).

Table 3.2. Summary of Graphene Electro-Optic Devices

Device	Electronic Fiber	Transistors	Tilted p-n
Uniqueness	Gate-controlled guided propagation	Large switching ratio, point source	Tilted gate, large switching ratio

3.3.1.1. Electronic Fiber

To spatially modulate the doping along the width in a graphene channel, a centrally overlaid top-gate adjacent to the direction of transport (length), and a combination of back-gate, under the whole graphene channel can be used.

The positive (negative) gate voltage on the back-gate and negative (positive) gate voltage on the top-gate forms an n-p-n (p-n-p) junction (Chauhan and Guo, 2011). It has been shown for such a device that total internal reflection at the p-n junction and electron waves undergo angle selective transmission, to be directed over the central region of graphene channel, for example, lasers pass through an optical fiber. The Fermi wavevector of the propagating electrons is controlled by the gate voltages and therefore the refractive index (η) (ratio of Fermi wavevector on two sides of the junction) which in result concludes the critical angle ($\theta_C = \arcsin(\eta)$). The substantial leakage to adjacent regions may have occurred for a large critical angle and reduced the guiding efficiency of the fiber. Apart from an optimal control of gate voltages, perpendicular magnetic fields can also be used to raise the incidence angle for an efficient guiding of electron wave over the central region, and in this manner lower the transmission across the n-p interface (Ouyang et al., 2007; Lam et al., 2010).

3.3.1.2. Transistors with High I_{on}/I_{off}

Extended drain and a dual-gate graphene transistor with a point source have been leading in Reference Sajjad et al. (2013), as shown in Figure 3.1(a). Using electro-optics as an operating principle, the design offers a subthreshold slope (SS) of less than $(K_B T/q) \log_e 10$ and an I_{on}/I_{off} ratio of five orders. The adjacent high transmission transverse modes and the Klein tunneling has been suppressed by advocating a physical insulating barrier else that would result in considerable leakage current. The angular bandwidth is modulated by the gate voltages over which by inducing a virtual energy band gap, from the graphene segment under V_{G1} to V_{G2} electrons can propagate, expressed by Eqn. (1). However, the performance of this design is comparable to a graphene quantum reflective switch which may be smooth to fabricate.

$$E_G = q(V_{G1} - V_{G2})\frac{2\sin\theta_B}{\cos^2\theta_B} \qquad (1)$$

3.3.1.3. Tilted p-n Junction Switches

Low-power logic devices have been presented to achieve the electro-optic effect in graphene. A dual goal is served by a tilted top-gate over the back-gated graphene. Besides the carrier doping is regulating in the graphene segment under it, a virtual energy band gap is administered by the inclination angle δ at the junction like barrier angle θ_B as expressed in Eqn. (1). The refractive index (η) is coordinated by the two gate voltages at the tilted junction and thus the critical angle (θ_C). An incidence angle of $\theta_i = \pi/2 - \delta$ is observed at the junction by the planar wavefront of electron waves from the source and can deteriorate total internal reflection if θ_i exceeds θ_C. Therefore, the electron waves can be reflected or transmitted by switching a suitable V_{G1} to switch the transistor (Schwierz, 2010; Taur and Ning, 2013).

3.3.2. Radio-Frequency Transistors

A switch-off state is entailed by only the large signal operation of a transistor. A transistor is biased at the quiescent point for a small signal operation, over which there is a superimposition of a small alternating component (ac). The saturation velocity and high mobility of graphene are congruous for such a small signal operation needed for microwave (μW) and radio frequency (RF) applications (Martin et al., 2008; Sarma et al., 2011).

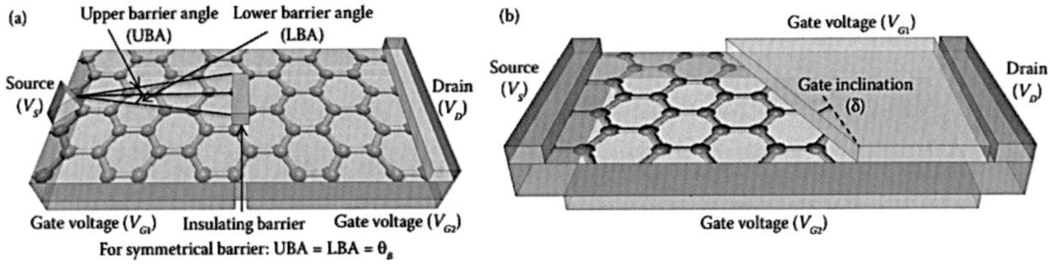

Figure 3.1. Electro-optics-based graphene switching device: (a) electrons are injected by point source overall angular modes (circular wavefront) into the channel; (b) at the tilted junction, the planar wavefront of electron waves injected from the source suffer refraction because of the electrostatically regulated refractive index through V_{G1} and V_{G2}.

Source: https://www.researchgate.net/publication/314459577_Applications_of_Graphene_Based_Materials_in_Electronic_Devices.

The intrinsic cut-off frequency (f_T) (>1 THz) has been absolutely demonstrated by the current progress in the fabrication of small channel length (L_g) RF GFETs, approaching that of mature III–V semiconductors. Furthermore, downscaling below the state-of-the-art 40 nm L_g GFET to sub-10 nm regime have been shown by simulations, that can drive f_T to few tens of THz. Additionally, as GFETs do not suffer from carrier freeze-out at cryogenic temperatures, hence as compared to other semiconductors, there is a larger thermal window for operation for RF performance in it. Thus, the recent hot research subject is tapping the immense potential of graphene RF technology and its challenges (Wu et al., 2011, 2012; Zheng et al., 2013).

High maximum-oscillation frequency f_{max} where unilateral power gain (U) for an impedance-matched system is one, and high cut-off frequency f_T where short-circuit current gain is unity, are the generally used figure-of-merit for an RF transistor where both of which are obtained by de-embedding S-parameters. However, good impedance matching and short-circuit output do not show a real system, for which a more practical criterion is open-circuit voltage gain (A_v) for comparing performance (Lin et al., 2009). These parameters for graphene RF transistors have been considerably improved by experiments, chiefly at Duan's Group (Los Angeles) and IBM Thomas J. Watson Research Center (Yorktown Heights). It is necessary to understand the factors that limit the performance for the appreciation of advancements from the initial demonstration of 26 GHz intrinsic f_T on mechanically exfoliated 150 nm long graphene channel to 300 GHz (350 GHz) on 40 nm long for CVD (epitaxial), 50 GHz extrinsic f_T on 170 nm long for CVD grown graphene, and 1 THz on sub-60 nm long for exfoliated, and to discern the unmet challenges. Particularly,

$A_v=g_m/(g_{DS}+ 1/R_{load}), f_T=g_m/2\pi C_{ox}$

and

$$f\max = fT/2\sqrt{(gDS(RGate + Ri + RS) + 2\pi f\, CT\, GDRGate)}$$

where; g_m represents transconductance; g_{DS} is the differential source/drain conductance; R_{load} indicates the load impedance during measurement; C_{ox} expresses the gate dielectric capacitance; R_{Gate} represents the gate resistance; R_i is the channel sided charging resistance of the gate/source capacitance; R_S is the source resistance; and C_{GD} is the gate-to-drain capacitance.

Especially keep that in mind that C_{ox} is directly proportional to length, g_m is inversely proportional (roughly independent) to length in the diffusive (ballistic) regime and directly proportional to mobility, which results in f_T a direct linear dependence on mobility and inverse second-order (first-order) dependence on length, showing that high mobility small channel length GFETs must have high f_T(Wang et al., 2010; Zhu et al., 2013; Han et al., 2014).

There is a significant degradation of the mobility of graphene as a result of the top dielectric and underlying substrate because of substrate-induced doping and their phonon modes. There is a relatively ignorable effect on g_m for the acoustic phonons in graphene itself. Hence, by migration from silicon dioxide (SiO_2) to boron-nitride (BN) dielectric and diamond-like carbon (DLC) substrate, RF performance has been improved. In addition to that, it is troublesome to accomplish ongoing saturation due to the deficiency of bandgap in graphene coupled with the substrate effect and so turns in g_m and an unstable quiescent point (Liao et al., 2010a, b, 2012). Nevertheless, current research has exhibited ongoing saturation with a moderate RF performance of GFET on a polyimide substrate (Farmer et al., 2009; Lam et al., 2012). From top-gate deposition-induced misalignment, the second level of f_T degradation occurs. It provokes overlap capacitance and parasitic-like large access resistance. With the use of the traditional complementary metal-oxide semiconductor (CMOS) technique and self-aligned gate structures with metal-T gates or nanowire (NW) gates by shadow masking, much enhancement has now been carried out. Later, the barrier in obtaining high f_{max} and A_v are the metallic contacts that form drain and source terminal (Table 3.3) (Meric et al., 2013; Zhu et al., 2015).

Table 3.3. Deterrents for Graphene RF Transistors

Problem	**Solution**	**Attained f_t or f_{max} (GHz)**
Bottom substrate and a top dielectric	Transition from SiO_2 substrate to BN dielectric and DLC substrate	f_T: 300 (I); 44 (I) (but highest f_{max}/f_T)
Gate misalignment induced parasitic	Self-aligned gates	f_T: 100–300 (I); 23 (E)
Absence of current saturation implying an unstable quiescent point	Polyimide substrate to accomplish current saturation	f_{max}: 10
Work-function difference at contacts' metal-graphene interface	Palladium contacts with full back-gating	f_{max}: 30

Note: E: Extrinsic; I: Intrinsic.

Because of the difference in the p-n junction produced between the graphene segments under channel and contacts, and work function (φ), charge carriers have to underpass through the electrostatic barrier make at the metal-graphene interface. Both A_v and f_{max} have been recently improved by setting the back-gate under the complete device to scale down the output conductance (g_{DS}) and to regulate the metal-induced doping in the graphene under contacts, and by the use of palladium electrodes that have lower φ mismatch (Farmer et al., 2010; Badmaev et al., 2012).

3.3.3. Thermal Transport Device

The thermal characteristics of graphene gain an increasing interest ignited by experiments exposed the superior thermal conductivity of graphene. It has been described that the thermal conductivity of the suspended single-layer graphene is in the range of around 3000–5000 W m^{-1} K^{-1} (Guerriero et al., 2012; Ramón et al., 2012), which shows its potential use as forthcoming thermal management material. Exceptionally high thermal conductivities have also been shown by the experiments for graphene flakes and few-layer graphene depending on the temperature and width (Gosh et al., 2008, 2010; Nika et al., 2009). Besides the high thermal conductivity, it has been shown by the investigations on transport properties of graphene high-field devices that the current situations and mobility are very sensitive to the dispersing of electrons by surface phonons of the SiO_2 substrate (Barreiro et al., 2009; Koh et al., 2010). Additionally, the substrate necessitates efficient heat dissipation for thermal interface material applications and low thermal boundary resistance at the interface of graphene. Therefore, the weak van der Waals interactions in out-of-plane thermal coupling owing to the very high anisotropy of thermal properties of graphene are overwhelmed by the vertically stacked graphene in a three-dimensional (3D) architecture (Dimitrakakis et al., 2008; Dorgan et al., 2010). The 3D nanostructures, e.g., graphene foams with a direct synthesis of 3D foam-like graphene macrostructures, nano-porous pillared graphene with combined forms of graphene sheets and CNTs, and 3D vertically aligned functionalized multilayer graphene architecture, have been demonstrated by the experimental and theoretical works which are valuable for flexible conductors, improved hydrogen storage, and energy management applications (Chen et al., 2011; Liang et al., 2011). Furthermore, thermoelectric power of ~50 μV K^{-1} at low temperature and ~80 μV K^{-1} at room temperature in high-mobility graphene samples have been demonstrated by the thermoelectric transport measurements of graphene (Zuev et al., 2009; Wei et al., 2009). Nonetheless, despite moderate thermoelectric power and high electronic conductivity, to attain an efficient thermoelectric energy conversion device, the thermal conductivity should be suppressed (Wang et al., 2009; Yang et al., 2010; Lin et al., 2011).

3.4. GNR DEVICES

3.4.1. Metal-Oxide-Semiconductor Field-Effect Transistor (MOSFET)

The most successful electronic devices are considered to be semiconductor transistors, particularly MOSFETs. Billions of transistors in our daily use devices are used for communication, computing, data storage applications, and display. In the earlier few decades, the struggle for larger density and higher speed transistor to improve their performance has favorably compelled the miniaturization of silicon planar MOSFETs. However, enduring this scaling trend, fundamental physical limitations will be faced by silicon-based MOSFETs. Hence, to meet future generation technology requirements as concluded by International Technology Roadmap for Semiconductors (ITRS), new channel materials with high carrier mobility are needed. As introduced earlier, extremely high mobility graphene is proposed to be the best candidate to meet this technology gap (Gargini, 2000).

Unfortunately, the absence of a bandgap in graphene restricts its relevance to traditional MOSFETs, which depend on bandgap to switch off the transistors to give a large ratio of on to off-state currents. Hence, there is a great interest in the AGNR as a potential candidate to reduce this problem due to its intrinsic semiconducting characteristic together with a tunable energy bandgap that relies on the atomic configurations and ribbon width at its edges. Especially, it is theoretically expected from the AGNR that it would show extremely high electron/hole mobility and promising electronic properties, identical to the characteristics observed in CNTs. Moreover, it has been shown that the upper-limit performance of ballistic GNR-FETs is identical to CNT-FETs, and must be able to outperform silicon MOSFETs in terms of 100% drive current abilities by the latest studies based on the semi-classical transport model coupled with band structure of nanoribbons using a single $_{Pz}$-orbital tight-binding method.

Furthermore, in terms of the nonequilibrium Green's function (NEGF) means, a 3D Poisson's solver was self-consistently coupled with more panoptic atomistic simulations based on quantum transport theory for the handling of electrostatics has been implemented to examine the current-voltage characteristics of GNR Schottky barrier (SB) FETs and GNR MOSFETs (Liang et al., 2007, 2008). A realistic real-space potential profile and the tunneling behavior can be captured by this approach that is overlooked in the conventional model. Withal, these play a significant part in the short channel effects of nanoscale FETs, e.g., drain-induced-barrier lowering (DIBL), and leakage current, induced by the direct tunneling from the source to the drain. It has been observed based on simulation outcomes that there might have potential in GNR MOSFETs with 1.5 nm width to outperform a double-gate 10-nm-scale Si MOSFETs in terms of DIBL, on-current, and SS. Still, as the performance of the GNR MOSFET-type device will be degraded by the tunneling, tunneling processes must bar the larger width GNRs from these applications. The inappropriateness of graphene with an infinite width for digital MOSFET applications is also indicated. However, other potential applications may be found by the high tunneling rate in GNR via direct bandgap, i.e., tunneling FETs or GNR MOSFETs operating in band-to-band tunneling (BTBT) mode.

Another critical view of the GNR FET device is that for the SB contacts, those metal should be collected in the source and drain region that might be expensive and challenging from the experimental and fabrication point of view. Whereas, graphene/doped graphene can be regarded as a metal with high mobility. Consequently, it is of great interest to understand the dependence of the device behavior of the GNR FETs on the shapes and contact types of the graphene sheet and as the devices' contacts have a full graphene device, so to operate the graphene sheet itself (Liang et al., 2007, 2008). Figure 3.2 shows the investigation of the effect of these five different types of contacts: (a) the semi-infinite normal metal with a constant surface density of states in the energy region of interest (L and $W \to \infty$), (b) the semi-infinite graphene sheet contacts (L and $W \to \infty$), (c) the finite-size rectangular graphene type of contacts, (d) the finite-size wedge-shaped contacts, and (e) the 1D zigzag GNRs contacts. Observe that the nanoribbons are nominated as armchairs by the shape of their edge or zigzag; it is in a sense contrary to the designation of CNTs.

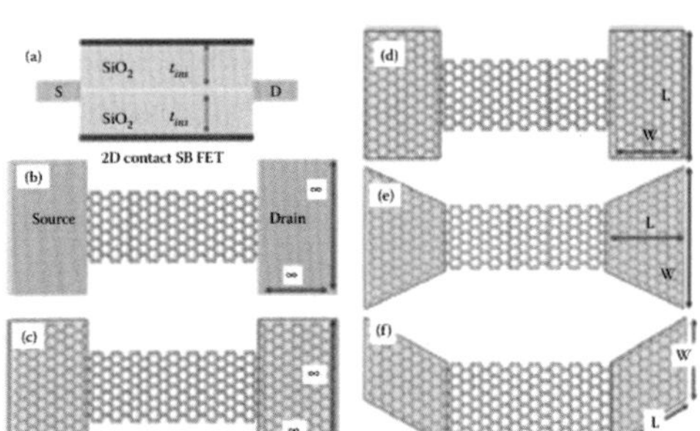

Figure 3.2. (a) Illustration of the simulated dual-gate graphene nanoribbon SB-FETs; (b) the drain and source parts are composed by a semi-infinite metal (L and W→∞); (c) the drain and source parts are composed by semi-infinite gra-

phene (L and W→∞); (d) the drain and source parts are composed by a finite rectangular shape graphene electrode; (e) the drain and source parts are wedge-shaped graphene sheets with zigzag edges; (f) the GNR contact with zigzag edges and same width as the channel.

Source: https://www.researchgate.net/publication/314459577_Applications_of_Graphene_Based_Materials_in_Electronic_Devices.

Poor performance has been observed for the intrinsic (undoped) semi-infinite graphene contacts as shown in Figure 3.3, but a better performance can be exhibited by the doped (electrostatically or chemically) graphene in terms of I_{on} to I_{off} ratio. In addition to that, the finite size rectangular-shaped graphene contacts and the finite size wedge-shaped graphene contacts exhibit better I_{on} to I_{off} ratios in contrast to the two former cases. The zigzag GNRs might be the best of these different graphene contacts that give promising I_{on} to I_{off} ratios, and also high I_{on} due to their strong metallic nature. It has been found as well that for interconnects, they could be the candidates between graphene or metallic electrodes and the channel (Areshkin and White, 2007).

All of the outcomes shown yet have been based on coherent transport. However, there is an important part of scattering mechanisms in realistic devices, in the carrier transport and generally decrease device performance. For example, the experimental results proposed that at high V_{DS}, the ballistic efficiency of I_{on} can come to 40% whereas, at low V_{DS}, I_{on} preserves about 15% of ballistic efficiency only. Mainly two scattering mechanisms, i.e., edge roughness (ER) (disorders) and phonon vibrations, can attribute this interesting phenomenon. The former encompasses optical and acoustic phonons that usually present in most materials at room temperature. It has been clear that there is a reduction of current for long channels by the acoustic phonon, still, its effect is ignorable in short-channel devices. Its mfp is approximately 10 nm for the optical phonon which even though of the order of the device channel being study, has less importance on influence on the device performance (Gupta et al., 2016). It is chiefly the high optical phonon energy of about 160 meV that turns into less uncertainty of backscattering in the device channel. In GNRs, ER is another important effect which has been recognized as the most challenging work from the fabrication aspect because of the complications in making perfect edges, although some groups have alleged to fabricate smooth-edged GNRs.

3.4.2. Tunneling Transistors

Although device performance might be enhanced and the device is kept successfully scaling down ultimately by novel channel materials, e.g., GNR, III–V materials, etc. It grows the power consumption in every chip. It has been anticipated that the power density in a high-end microprocessor can early arrive at the level of a nuclear reactor and indeed the level of the sun's surface in decades, a condition which is certainly indefensible (Yoon and Guo, 2007).

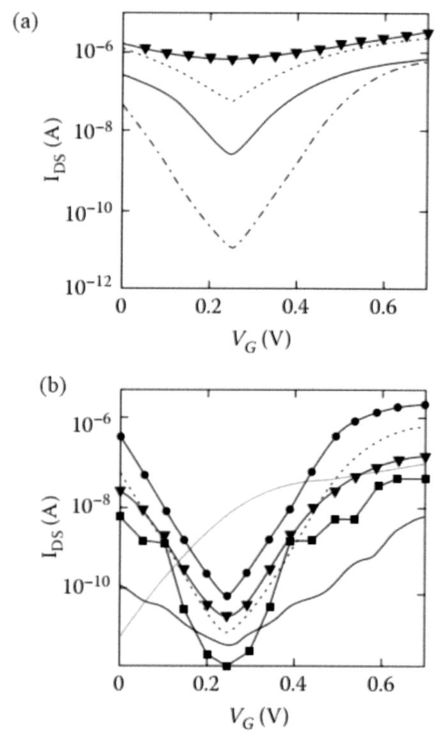

Figure 3.3. (a) Transfer properties of the armchair GNR SBFET with widths of 4 nm (triangle), 1.6 nm (dashed), 1.4 nm (solid), and 3 nm (dot-dashed) under VDS = 0.5 V.; (b) Transfer characteristics of the 1.6 nm wide GNRSBFET with different types of graphene type of contacts: the semiinfnite normal metal (dashed line), the semi-infnite graphene sheet, fnite sizenrectangular shape graphene sheet (L = 3 nm, W = 5 nm), wedge shape graphene sheet (L = 3 nm, W =5 nm), and zigzag GNR contacts(L = 3 nm, W = 1.6 nm)

Source: https://pubs.acs.org/doi/abs/10.1021/nl080255r.

The reason is that, as the increase in the number of transistors per chip, the power density has been increasing exponentially because of consumed heat in chips. Unluckily, this issue may not be resolved with traditional CMOS technology on account of power dissipation from a constitutional limit of the typical operation principle of MOSFETs by the reason of its reliance on thermionic electron transport, i.e., 60 mV/decade of the SS at room temperature. This thermionic mechanism confines the capability to downscale the operating bias tends to high power consumption in MOSFETs (Leong et al., 2009; Lam et al., 2010a, b).

Consequently, to scale down the power consumption and bias of electronic devices, novel device structures based on various carrier transport mechanisms and device physics have been suggested. A lot of attention has been attracted by TFETs from the device engineering community among various devices proposed because of their device structure which is identical to MOSFETs, and their simple operation principle, the latterly being of greater concern for the semiconductor industry.

The TFET is a three-terminal p-i-n device, unlike MOSFETs. As shown in Figure 3.4, the off-state current of TFETs is regulated by the electron tunneling through the source to the drain. The current is suppressed because of the long tunneling path and must be very low. When the current rapidly increases because of a strong electrical field between the tunneling junctions and the short tunneling path, the channel potential moves down because of the applied positive gate bias, until the alignment of the top of the valence band edge of the source and the band edge of the channel conduct. It is known to be the on-state current. After all, this device is not depending on the thermionic electrons to transit via a barrier, the operation bias can be a few 100 mV in principle, and small, contributing to a considerable reduction in the power devouring of each device.

Even though Si-based TFETs have been discussed earlier, their on-state current is extremely small due to their indirect bandgap and heavy effective mass, which portrays them unfit for practically. Moreover, a few other enhanced structures have also been proposed, e.g., III–V TFETs, broken gap TFETs, and Ge TFETs, to give a higher on-state current because of their direct bandgap, smaller effective mass, and a sharper electrical field between the drain and channel. For TFETs, in addition to these materials, AGNR has also been treated as a potential candidate due to the tunability of the bandgap by regulating the GNR's width and the smaller effective mass of the carrier. Subsequently, a heterojunction can be fabricated with the same materials to bypass lattice mismatch from peculiar semiconductor materials.

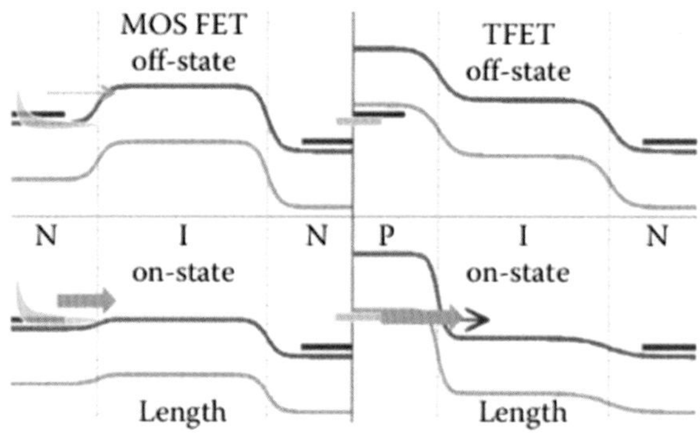

Figure 3.4. Representation of the operation principle of MOSFETs and TFETs.

Source: https://www.researchgate.net/publication/314459577_Applications_of_Graphene_Based_Materials_in_Electronic_Devices.

It has been shown by the earlier theoretical studies that an on to off current ratio of more than 10^{10} can be reached by GNR TFETs with the SS can be <10 mV/decade and a gate voltage sweep of only 300 mV. In addition, generally, it is found that AGNR tunneling FETs from the 3p + 1 family exceed the other two groups (3p and 3p + 2), in the context of larger I_{on} to I_{off} ratio due to smaller effective mass of electrons in 3p + 1 AGNR. The doping concentration at the source has a significant role to optimize a TFET. With its increases, I_{on} also increases for the sake of short channel length and the higher electrical field; however, there is a degradation of subthreshold swing because the thermionic electrons increase leads to the degradation of the off-state current. Moreover, in contrast to ITRS requirements, promising scaling behaviors can be indicated by the power-delay and intrinsic delay product of the GNR tunneling FETs to meet future requirements (Yoon and Salahuddin, 2010).

Another auspicious characteristic of GNRs is their feasible geometry change from CNT and a tunable bandgap to scheme an interface strain engineering, heterojunction, and partially unzipped CNT-GNR interface, coming from various widths. Numerous theoretical proposals have been proposed of enhancing tunneling current in TFETs, based on these fabrication technologies (Lu and Guo, 2010). For instance, it has been demonstrated that with the use of the heterostructure to add a large E_g material in the middle and a small E_g material in the channel near the source reduces I_{off} and improves I_{on} leads to a larger I_{on}/I_{off}. For device scaling, such a device structure is promising as well. As the channel length reduces to 10 nm, the theoretical simulation predicts I_{on}/I_{off} to four orders of magnitude and I_{on} to reach 1.5 mA μm^{-1}.

3.4.3. Resonant Tunneling Devices

Resonant tunneling diode (RTD) is another attractive device regulated in tunneling mode, which is predicted to have good potential for high-speed and low-power applications (Luisier and Klimeck, 2009). The RTD is a two-terminal device dissimilar to a FET device, and its operation principle is based on the quantum states formed between two wells to give resonant tunneling, and a semiconductor source to avoid the transportation of electron because these resonant states collapse onto the bandgap region of the source (Sun et al., 1998; Akeyoshi et al., 1999). Tunneling via two barriers is started by electrons when the bias is applied, with short tunneling probabilities, and after that, the current increases hastily due to high transmission rate devoted by the resonant states, so-called peak current (state) since the two wells drop into the electron sea of the source generates the quantum states. Nonetheless, the current drops as the state keep dropping below the CB edge of the contacts, because according to coherent transport, no electron in the bandgap region of the source can input into the state. Hence, the current only depends on the small tunneling probability and by standard, remains small, known as valley current (state). Subsequently, a conductance or negative resistance can be produced between the valley and peak states, which yields two distinguishable states for circuit design.

Commonly, RTDs comprises of heterojunctions of various semiconductor materials, i.e., the InGaAs/AlAs double-barrier structure. Unfortunately, dislocations, mismatched lattice, and phonon scattering reduce the peak-to-valley ratio (PVR) and hinder development and progress. Whereas, as recommended earlier, numerous widths of GNRs are able to effectuate various E_g that gives the needed double-barrier structures from a single material for RTDs. Attractively, as illustrated in Figure 3.5, to accomplish the same electronic requirement, devices of different shapes (H, S, and W shape) can be made as well. The latest research that is based on the nearest-neighbor-orbital tight-binding coupled together with quantum transport simulations demonstrates that all of these three different-shaped RTDs express negative differential resistance (NDR) properties and likely the peak state position represents that the robust approach of RTD operation is chiefly dependent on its double-barrier structure and to a minor extent on the geometrical shapes. Still, there is a significant role of the geometrical shapes on the magnitude of the valley and the peak current. For instance, it is noted that the largest I_{peak} is provided by the W-shaped AGNR RTD but a higher PVR was offered by H-shaped AGNR RTD. The interface scattering can associate the observation for various shapes. There are direct paths for the W- and S-shaped cases to permit an electron to pass through; on the other hand, more interface variation is faced by the electrons in the H shape leading to the suppression of the transmission rate. Fortunately, it also helps to obstruct the current in the valley states, and hence, increases the PVR.

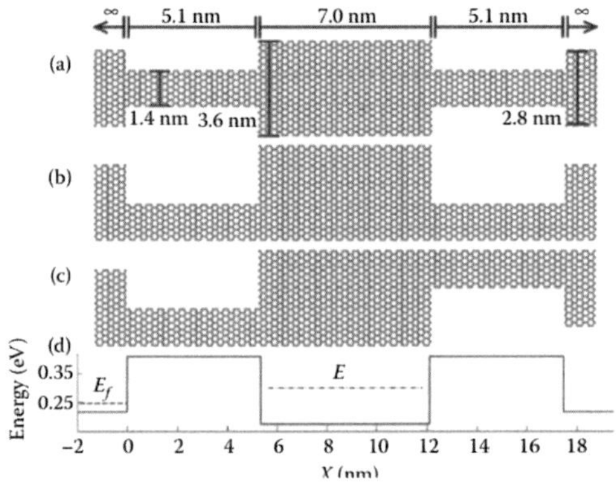

Figure 3.5. Illustration of the simulated AGNR RTDs with various shapes, i.e., (a) H; (b) W; and (c) S shapes. Energy band diagram for these differently shaped AGNR RTDs at equilibrium is demonstrated in (d), showing the DBQW structure for RTD operation.

Source: https://www.researchgate.net/figure/Schematics-of-the-simulated-AGNR-RTDs-with-different-shapes-namely-a-H-b-W-and-c_fig2_314459577.

Nonetheless, akin to the TFETs and MOSFETs, ER undoubtedly affects the performance of GNR RTDs. As stated by a theoretical study, the electronic structures will be upset by the ER leading to a nonuniform barrier greater than the well and broadening of the resonant states (Liang et al., 2010). Moreover, defect states within the wells are constructed. The tail of the wave function of the electrons in the contact to increase the valley current markedly and to infiltrate deeper into the quantum well region is allowed by both effects, turning to the degradation of the PVR.

3.4.4. Interconnects

The transistors have to be unified into functional blocks that additionally connect with other similar sections on an integrated circuit (IC). The clock, data, ground signal, and power are routed among these using interconnects of differing bandwidth, dimension, and current-carrying ability based on the design, or a wiring system (Areshkin and White, 2007; Behnam et al., 2012). Carbon-based materials, e.g., GNR, and CNT are expected by ITRS to be a substitute for conventional copper interconnects because of the latter's immense resistivity for small wire widths, due to electromigration effects, and grain boundary scattering for lesser technology nodes (beyond 16 nm). An edge over CNT is given by the fabrication ease of GNR for the formation in devices. The management of large current densities (>10^9 A cm^{-2}) explicit of material breakdown is enabled by the powerful sp^2-hybridized σ-bonds in GNR (Xu et al., 2009; Sarkar et al., 2011). Moreover, large mfp (~1 μm), low in-plane dielectric constant (κ), and large thermal conductivity that pacifies electromigration, thus for high-frequency chips, high momentum relaxation time arranges GNR as a good interconnect material. Still, there is a low mode density of monolayer GNR interconnect which makes causes it highly resistive, whereas interlayer coupling in multilayer GNR (MLGNR) scales down electron mobility (Naeemi and Meindl, 2007; Das and Rahaman, 2012). The intercalated zigzag-edged MLGNR (side contacts) of high specularity (p ~ 1), i.e., inducing roughly with no momentum change with the length of ribbon because of edge scattering of enough interlayer separation is so treated interconnects that carry high and low frequency signals and an imminent solution for power interconnects. Observe that only doped zigzag MLGNR is referred to be the consequent discussion on GNR for interconnects unless stated else (Bolotin et al., 2008; Santos and Kaxiras, 2013).

Figure 3.6 illustrates a physics-based model of a GNR interconnect. It has been demonstrated with the help of this model that as compared to Cu local (global) interconnects at 11 nm node, stage-2 AsF_5 doping in MLGNR of moderate specularity (p= 0.8) gives much better performance, even though SWCNT improves MLGNR for global interconnects (Naeemi and Meindl, 2008; Haji et al., 2012). The Nyquist stability of the MLGNR interconnect and transient analysis of step response with finite contact resistance using the same model, additionally exhibits that it has marginally steady overshoots and insignificant stability, which declines as its length and width decrease even though the propagation delays are less than SWCNT. Hence, MLGNR must capable of speedy signal propagation as a local interconnect if the overshoots are within the design constraints. Afterward, for global interconnects, the wire width can be of the order of 1 μm which is extremely greater than the skin depth at high frequencies (>1 GHz). The current density in GNR is redistributed by the resultant anomalous skin effect (ASE) that has to be constructed by self-consistently solving the Boltzmann equation using a nonlocal electric field in the wire. The resistance of MLGNR is estimated to be less than the resistance of the SWCNT bundle (Cu) up to 200 GHz (entire μW range) even of low specularity by the comprehensive simulations (Sarkar et al., 2011; Haji et al., 2012).

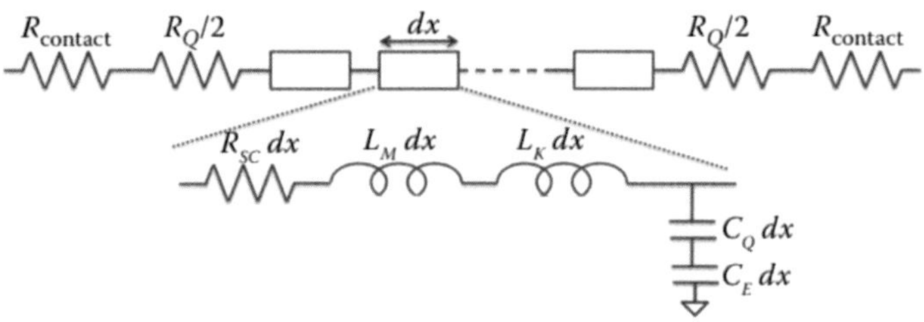

Figure 3.6. An illustration of equivalent circuit of transmission line of GNR interconnects wire

Source: https://aip.scitation.org/doi/10.1063/1.3624459.

Two key differences are noticed between the side-contact MLGNR model and the experiments explained above. The Fermi energy is shifted from the neutral point by the substrate-induced doping; however, the effect reduces exponentially toward top layers (Naeemi and Meindl, 2009). The disparate conductance of each GNR layer leads to divergence from the consistently conducting independent layer model. Next, in experiments, on top of the MLGNR, contacts are connected; thus, opposite to the supposition that conductance grows with the number of GNR layers, conductance has currently been exposed to saturate with layers as the contacts efficiently couple only with petty top layers. The energy-delay product (EDP) and signal delay are reduced for an optimal number of layers, which is an operation of interlayer coupling, specularity, and interconnect length. Still, the performance of top-contacted MLGNR interconnects with many smooth edges might be better than Cu for short lengths, although the performance can be upgraded even for longer interconnects with doping optimizations (Kumar et al., 2012). It must be kept in mind that MLGNR may change to graphite about eight layers of thickness tearing it useless for interconnects. An upper boundary is put on the choice of an optimized number of layers evaluated from the analytical model. It is referred to Rakheja et al. (2013) for a more comprehensive review on performance evaluation of GNR interconnects counter to copper/low-κ interconnects (Nirmalraj et al., 2011; Jain et al., 2012).

3.5. SPINTRONIC DEVICES

A rich set of properties conducive for spintronic devices is manifested by graphene. High thermal conductivity, bandgap tunability, and long spin-diffusion length has schemed for possibilities of the new device as explained in subsections (Table 3.4) (Nishad and Sharma, 2013).

Table 3.4. Summary of Graphene-Spintronic Devices

Spintronics Devices	Features	Challenges
Magnetoresistance device	1. The MR ratio of GNRs is stable against edge roughness 2. Conductance in GNRs and tunability of bandgap by magnetic fields	Wide GNRs are needed to have a moderate band gap modulation with practical magnetic fields at room temperature
Spin diode	Highly spin-polarized currents are allowed to be generated and tuned by source-drain voltages or/and magnetic configurations by the transmission selection rule in even-N ZGNRs	Magnetized ZGNRs with smooth edges are essential
Spin-logic device	A complete set of logic gates (AND, NAND, OR, and NOR) can be designed using the above spin diode	–
Spin transistor	1. Datta-type spin transistor by proposing spin-orbital coupling into the graphene sheet as a result of the hybridization effect with heavy 5d-elements or the proximity effect using a ferromagnetic insulator. 2. BJT- or FET-type spin transistors used the edge states of ZGNRs	Metallic states near the Fermi level and weak spin-orbital coupling because of the hybridization or proximity effect ZGNRs with smooth edges are essential
Spin caloritronic device	A temperature difference between the source and drain can drive the current flowing through magnetized ZGNRs	–

3.5.1. Magnetoresistance (MR) Device

The $n= 0$ Landau level has been predicted by the theoretical studies in the existence of a very high magnetic field, a characteristic one of a kind to graphene-based materials. Because it is anticipated from theoretical studies of this phenomenon, a fall in the bandgap of GNR with the rise in the magnetic field has been successfully reported by a recent experimental study, even though observance of this phenomenon is challenging for practical magnitudes of a magnetic field (Bai et al., 2010; Nirmalraj et al., 2011). With respect to this interesting property, the magnetic field can modulate the electron conductance turning to magnetoresistive effects in GNR devices. A very large magnetoresistance (MR) has been illustrated tentatively. Opposite to traditional spin-valves that require ferromagnetic lead, the advantage of this device is that this MR just needs full graphene (GNR) sheet and is the intrinsic material behavior, which must be straightforward for industrial fabrication. In addition, in contrast to TFETs, RTDs, and MOSFETs where ER decreases the performance typically, the impact of ER is comparatively less for GNR MR-devices because the Lorentz force limits the electron to the edge that leads to the high transmission rate. A theoretical study, however, has also observed that a large width GNR with a small bandgap is essential to have a bandgap modulation for smaller magnetic fields (Kumar et al., 2010). The transport properties can be taken over by the thermionic electrons and suppress MR at a higher temperature because of the dissemination of the Fermi distribution function, which is indicated by a small bandgap. Therefore, a p-i-n heterostructure AGNR device has been theoretically presented to deceive this problem. In this structure, a narrow GNR is implemented in the contact region, whereas a wide GNR is implemented in the channel region. Similar to tunneling FETs, the operation of the former is to hinder the thermionic electrons in the source because of its bandgap and function the larger width GNR in the channel within a small magnetic field. Furthermore, due to the varying size of contacts and channel, to improve the current to prompt other devices in the circuit, multiple GNR strips can be carried out as contacts. As illustrated in Figure 3.7, it has been observed that three orders under 5 T of field can be reached by the MR of this p-i-n heterostructure AGNR device as compared to about 10% MR using normal metal contacts (Kumar et al., 2010).

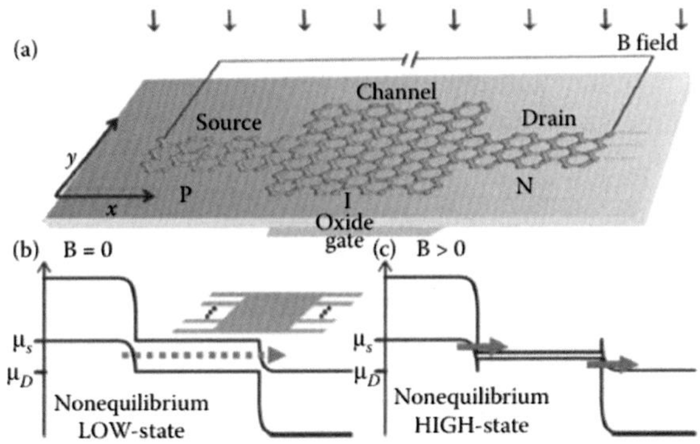

Figure 3.7. Illustration of a ZGNR-based bipolar spin diode. ML, R shows the magnetization of the electrodes, whose values can be ±1 and 0 which correspond to magnetization along ±y direction and nonmagnetic lead, respectively. (a, b) For [ML,MR] = [1,0], a positive (negative) bias can only drive spin-down (up) electrons over the device.

Source: https://www.researchgate.net/figure/Schematic-illustrations-of-ZGNR-based-BJT-type-transistors-a-c-Top-and-side-view-of_fig5_314459577.

3.5.2. Spin Diode

In addition to spin filter and half-metallic effects, ZGNRs have been expected to show MR effect, which is necessary for graphene-based magnetic storage applications. A very large MR can be attained in a ZGNRs-based spin valve is the prediction of Kim et al. (2008). Their first finding is that a weaker magnetic field can be utilized for a ZGNR of larger width; and a moderate magnetic field >2.2 T is abundant to transit a wide ZGNR with 8.73 nm width, for its ground state to a magnetized state (Tombros et al., 2007).

The MR ratio is calculated as $MR= (R_{AP}–R_P)/R_P$, where R_P and R_{AP} are the resistances in the antiparallel configurations and magnetic parallel, respectively. At room temperature, the calculated MR ratio can be as tremendous as 10^6 for an 8-ZGNR. The large MR has participated in the transmission selection rule as the result of the mismatching/matching wave function symmetry in spin sub-bands. The next design used the transmission selection rule in a spin diode demonstrated in Figure 3.8, in which magnetic configurations and/or the source-drain voltages can generate and tune highly spin-polarized currents, which is summed up in Figure 3.9 (Liang et al., 2011).

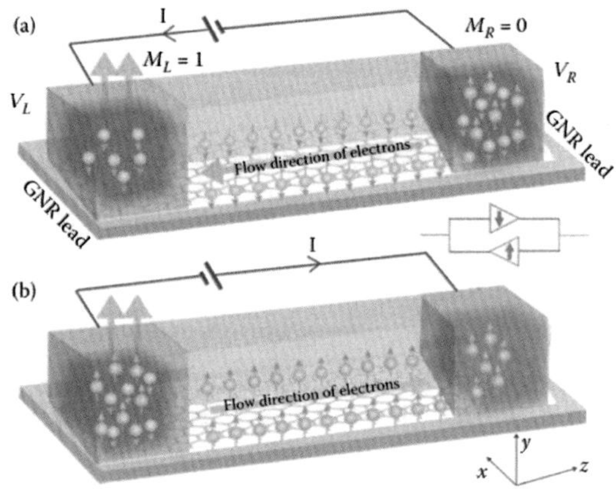

Figure 3.8. A spin current flowing via the ZGNR channel can be chosen by magnetic configurations and source-drain voltages.

Source: https://journals.aps.org/prb/abstract/10.1103/PhysRevB.83.115427.

$[M_L, M_R]$ / Bias	[1, 1]	[1, −1]	[1, 0]	[1, 0]	[0, 0]
+	↑↓	↓	↓	↑	↑↓
−	↑↓	↑	↑	↓	↑↓

Figure 3.9. Diagrammatically view of the spin NOT gates. The input is labeled as A, whereas the output terminal is labeled as Y. A magnetization-fixed terminal (Mref) is used as a reference. The logic input and output 1 (0) are encoded by the magnetization 1 (–1) of the input terminals and the output current for 1 (0) including (excluding) the spin-up current, respectively.

Source: https://journals.aps.org/prb/abstract/10.1103/PhysRevB.83.115427.

3.5.3. Spin-Logic Device

The construction of spin logic devices is possible by ZGNRs-based spin diodes. The magnetization of the terminals encodes logic inputs. If the output current includes (excludes) the spin-up current, then the logic output is encoded to be 1 (0). The illustration of a NOT gate is shown in Figure

3.10. The spin-up channel is conducting, and corresponding to a logic output 1, when the magnetization of the left electrode is fixed to –1 (logic input 0). Whereas, if the magnetization of the left electrode is fixed to 1 (logic input 1), then the spin-down channel is conducting (logic output 0). Thus, the NOT logic operation is performed. Likewise, other logic gates can be realized as well based on the above-given design concepts (e.g., OR, NOR, AND, and NAND gates). Further designs of complex spin logic operations are allowed by these logic gates and cover the way for the full deployment of spintronic computing devices (Zeng et al., 2011).

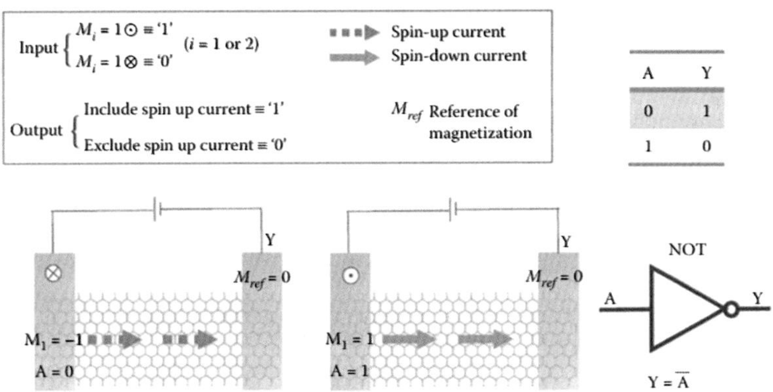

Figure 3.10. Diagrammatic illustrations of NOT gates. The input and output termini are labelled as A and Y, respectively. A magnetization-fixed terminal (Mref) is employed as a reference. The logic input and output 1 (0) are encoded by the magnetization 1 (-1) of the input points and the output current for 1 (0).

Source: https://journals.aps.org/prb/abstract/10.1103/PhysRevB.83.115427.

3.5.4. Spin Transistor

For the Datta-type spin FET, great attention for the consumption of graphene as a channel has been attracted by the large spin relaxation time and long spin diffusion length in graphene. Nonetheless, two critical issues must be resolved. A high-quality tunnel barrier and ferromagnetic metals are required by detection at the source and drain and an efficient spin injection by the reason of the fact that graphene is nonmagnetic. Additionally, for designing Datta-type spin FETs, there is a strong limitation, i.e., the ignorable spin-orbit coupling (SOC) in graphene. This limitation may overcome with a ferromagnetic insulator (FMI) as the gate dielectric. The initiating from exchange interactions between graphene is affected by the electric-field-controlled proximity and a large spin splitting can be induced by FMI at graphene. A large SOC was induced by this externally, and has also been noted when Au is intercalated at the graphene-Ni interface due to the hybridization of Dirac electrons with Au 5d states in graphene. With the exploration of the unique edge states in ZGNRs, new kinds of spin FETs can be designed. In comparison with Datta-type spin FETs, the key benefit of these designs is to exclude the demand for ferromagnetic metals and SOC for detection and spin injection (Shishir and Ferry, 2009; Liao and Duan, 2012; Soonmin, 2019).

3.5.4.1. Bipolar Junction Transistor (BJT) Type

Three terminal spin transistors can be designed with the help of ZGNR-based spin diodes as building blocks. Two designs of the transistor are shown, Figure 3.11(a) shows for spin-up current, while Figure 3.11(b) shows for spin-down current. The side view and equivalent circuit diagrams of the device are illustrated in Figure 3.11(c) over e, respectively. Each transistor consists of an emitter, a base, a collector, and three terminals. The emitter is grounded, the voltage of the base (V_B) can be fluctuated which regulates the polarization and flow of the current, and the collector voltage (V_C) is set to a positive value. The magnetic configuration within the collector and the emitter determines the polarity of the spin transistor. As demonstrated in Figure 3.11(f), when $V_B/V_C = ½$, the current gain (I_C/I_B) rises dramatically, representing a high amplification for spin-polarized currents with an appropriate bias condition. Transistors operating as voltage amplifiers can be constructed in the same manner (Low and Appenzeller, 2009; Sohier and Yu, 2011).

The design of a Johnson-type voltage amplifier is shown in Figure 3.11(g) and h. Hither, to detect the spin-polarized current, a ferromagnetic cobalt electrode is used as the collector. Under a bias voltage, the spin-polarized current flows from the emitter to the base and after that diffuse from the base to the collector, producing a voltage difference that can be evaluated. It is possible for the transistor to serve as a voltage amplifier with a decent selection of parameters (Gupta et al., 2015; Sajjad et al., 2013).

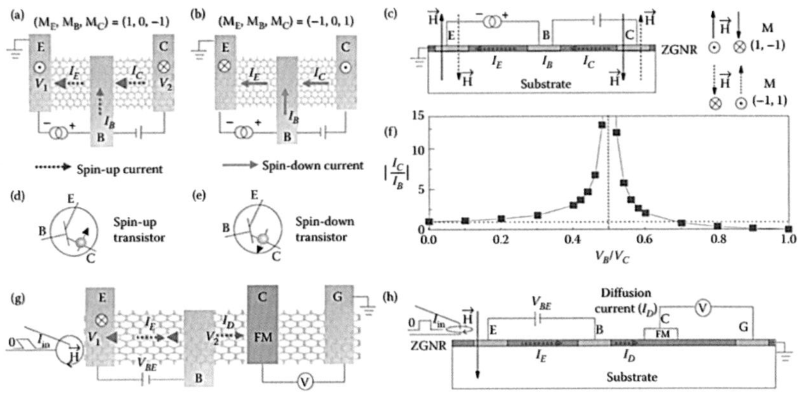

Figure 3.11. Diagrammatic illustrations of ZGNR-based BJT-type transistors. (a)–(c) Top and side view of a spin current amplifier. Their circuit symbols are displayed in (d, e). (f) Current gain (|IC/IB|) as a function of (VB/VC). (g, h) Top and side views of a Johnson-type spin voltage amplifier.

Source: https://journals.aps.org/prb/abstract/10.1103/PhysRevB.83.115427.

3.5.4.2. FET Type

Despite the above described bipolar-like transistors, A ZGNR-based spin FET is proposed by Guo et al. (2008) in which a transverse gate electric field controls the conductance of source-drain. The proposed device comprises a transport channel that is adjusted by two pairs of electrodes. The voltages on the left pair of electrodes are absolute, and the conductance of source-drain is managed by a transverse electric field producing by the right pair of electrodes. The direction and value of the transverse electric field can control the transmission of spin-down or spin-up electrons. If the transverse field is ranging from –0.1 to +0.1 V/A, 10^3 can be reached by the on/off ratio (Cheianov and Fal'ko, 2006; Sutar et al., 2012).

3.5.5. Spin Caloritronic Device

In ZGNRs, spin-polarized currents can be produced with a thermal bias even without an external bias voltage. Figure 3.12(a) shows the diagrammatic schematic illustration of a ZGNR-based spin caloritronic device. A temperature difference (T_{SD}), the difference between the source temperature (T_S) and the drain temperature (T_D), i.e., T_S–T_D, drives the current flowing through the magnetized ZGNR (M-ZGNR). As illustrated in Figure 3.12(b), opposite flowing spin-down and spin-up currents can be produced in M-ZGNRs with a temperature difference. In M-ZGNR, the asymmetric transmission spectra of spin-down and spin-up electrons associates this appealing spin Seebeck effect. Moreover, a back-gate voltage can be enforced to deal with the spin currents illustrated in Figure 3.12(c). As T_{SD} increases and V_G is fixed, spin currents (I_{SD}) increase and attain saturation signifying a thermal spin transistor. Furthermore, for a fixed T_S and T_{SD}, tuning a back-gate voltage achieves the spin currents with full polarization (Williams et al., 2011; Wu et al., 2011). Moreover, as demonstrated in Figure 3.12(d), when the magnetic state of ZGNRs is changed from the ferromagnetic state to its ground state, a very large thermal MR effect is found. To identify the electronic states of ZGNRs, a unique technique is provided by the sensitivity of thermally induced spin currents to the magnetism of ZGNRs. In M-ZGNR, a rich set of thermal spin factors (spin diode, spin filter, MR device, and spin FET) are provided by the flexible control over thermally induced spin-polarized current, by electrical (gate voltage), thermal (i.e., temperature), and magnetic means. These spin factors assist in the dissipation of heat in a working nanoelectronic device and open the gate to carbon-based spin caloritronics applications. Despite the M-ZGNRs, in spin valves and ZGNRs-based heterostructures, the impact of thermally induced spin currents has also been examined. For creating thermally induced spin currents, the reason for the ZGNRs system to be a good applicant is the fact that its low-energy electronic states are sensitive to the surroundings, e.g., external magnetic, different edge terminations, or electric fields, which shatters the electron-hole symmetry and increases to thermal-induced spin currents (Banerjee et al., 2008; Sajjad and Ghosh, 2011).

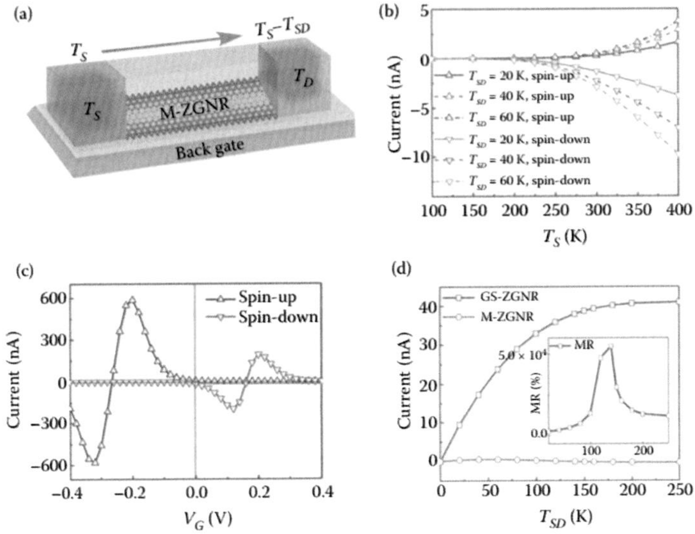

Figure 3.12. Diagrammatic of thermally induced spin currents in (a) M-ZGNR. Spin-down currents in the M-ZGNR (spin Seebeck effects) and opposite-flowing spin-up can be induced by a temperature bias. (b) For different TSD, spin-dependent currents as a function of TS. (c) Gate-dependent spin-polarized currents (TS = 400 K, TSD = 60 K). (d) Spin currents as a function of TSD for GS-ZGNR and M-ZGNR (TS = 400 K, VG = –0.02 V). The inset indicates MR can be as high as 5 × 104% by translating ZGNRs from ferromagnetic to ground state.

Source: https://pubs.rsc.org/en/content/articlelanding/2013/nr/c2nr32226a.

Furthermore, it is found that in a ZGNRs-based spin valve, the local magnetic moment over a magnetic atom is vigorously associated with the thermal bias. An effective strategy to store information in a spin valve is provided by this thermally controlled magnetic distribution. As shown in Figure 3.13, the ZGNR-based spin valve is in the antiparallel configuration and with a Ni atom absorbed on one edge of the ZGNR. In the ZGNR, the absorbed Ni atom for T_{SD}= 0 drives a magnetic moment of almost –0.01 μ_B. This figure is much low than that of an isolated Ni atom, which can be as high as 5.6 μ_B. In the ZGNR, the quenching of the Ni magnetic moment expresses the charge transfer (CT), or solid coupling, between the Ni atom and its environment. Interestingly, as exhibited in Figure 3.13(b), the magnetic moment of the Ni atom can be markedly restored by the thermally induced current, which can be upgraded highly to 1 μ_B by practicing a thermal bias of T_{SD}= 40 K. Thus, besides the firm coupling between the Ni atom and its environment, the Ni magnetic moment can be partially aligned to its original value by the injected spin currents (Alam, 2009; Xia et al., 2010).

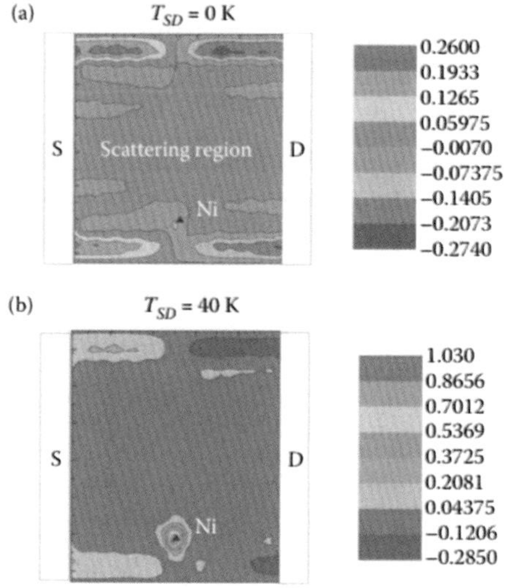

Figure 3.13. The thermal-controlled magnetic moment of a magnetic impurity in a ZGNR-based spin valve. (a, b) The spatial magnetic moment in the scattering region for TSD = 0 and 40 K, respectively. The magnetic moments of the Ni atom in (a) is–0.01 μB and (b) is 1 μB.

Source: https://pubs.rsc.org/en/content/articlelanding/2013/nr/c2nr32226a.

REFERENCES

1. Akeyoshi, T., Matsuzaki, H., Itoh, T., Waho, T., Osaka, J., & Yamamoto, M., (1999). Applications of resonant-tunneling diodes to high-speed digital ICs. In: *Conference Proceedings. Eleventh International Conference on Indium Phosphide and Related Materials (IPRM'99) (Cat. No. 99CH36362)* (pp. 405–410). IEEE.
2. Alam, K., (2009). Uniaxial strain effects on the performance of a ballistic top-gate graphene nanoribbon on insulator transistor. *IEEE Transactions on Nanotechnology, 8*(4), 528–534.
3. Areshkin, D. A., & White, C. T., (2007). Building blocks for integrated graphene circuits. *Nano Letters, 7*(11), 3253–3259.
4. Areshkin, D. A., Gunlycke, D., & White, C. T., (2007). Ballistic transport in graphene nanostrips in the presence of disorder: Importance of edge effects. *Nano Letters, 7*(1), 204–210.
5. Badmaev, A., Che, Y., Li, Z., Wang, C., & Zhou, C., (2012). Self-aligned fabrication of graphene RF transistors with T-shaped gate. *ACS Nano, 6*(4), 3371–3376.
6. Bae, S., Kim, H., Lee, Y., Xu, X., Park, J. S., Zheng, Y., & Kim, Y. J., (2010). Roll-to-roll production of 30-inch graphene films for transparent electrodes. *Nature Nanotechnology, 5*(8), 574.
7. Bai, J., Cheng, R., Xiu, F., Liao, L., Wang, M., Shailos, A., & Duan, X., (2010). Very large magnetoresistance in graphene nanoribbons. *Nature Nanotechnology, 5*(9), 655–659.
8. Balandin, A. A., Ghosh, S., Bao, W., Calizo, I., Teweldebrhan, D., Miao, F., & Lau, C. N., (2008). Superior thermal conductivity of single-layer graphene. *Nano Letters, 8*(3), 902–907.
9. Banerjee, S. K., Register, L. F., Tutuc, E., Reddy, D., & MacDonald, A. H., (2008). Bilayer pseudospin field-effect transistor (BiSFET): A proposed new logic device. *IEEE Electron Device Letters, 30*(2), 158–160.
10. Barreiro, A., Lazzeri, M., Moser, J., Mauri, F., & Bachtold, A., (2009). Transport properties of graphene in the high-current limit. *Physical Review Letters, 103*(7), 076601.
11. Behnam, A., Lyons, A. S., Bae, M. H., Chow, E. K., Islam, S., Neumann, C. M., & Pop, E., (2012). Transport in nanoribbon interconnects obtained from graphene grown by chemical vapor deposition. *Nano Letters, 12*(9), 4424–4430.

12. Bolotin, K. I., Sikes, K. J., Hone, J., Stormer, H. L., & Kim, P., (2008). Temperature-dependent transport in suspended graphene. *Physical Review Letters, 101*(9), 096802.

13. Bolotin, K. I., Sikes, K. J., Jiang, Z., Klima, M., Fudenberg, G., Hone, J. E., & Stormer, H. L., (2008). Ultrahigh electron mobility in suspended graphene. *Solid State Communications, 146*(9, 10), 351–355.

14. Chauhan, J., & Guo, J., (2011). Assessment of high-frequency performance limits of graphene field-effect transistors. *Nano Research, 4*(6), 571–579.

15. Cheianov, V. V., & Fal'ko, V. I., (2006). Selective transmission of Dirac electrons and ballistic magnetoresistance of n-p junctions in graphene. *Physical Review B, 74*(4), 041403.

16. Cheianov, V. V., Fal'ko, V., & Altshuler, B. L., (2007). The focusing of electron flow and a Veselago lens in graphene p-n junctions. *Science, 315*(5816), 1252–1255.

17. Chen, Z., Ren, W., Gao, L., Liu, B., Pei, S., & Cheng, H. M., (2011). Three-dimensional flexible and conductive interconnected graphene networks grown by chemical vapor deposition. *Nature Materials, 10*(6), 424–428.

18. Chin, S. K., Seah, D., Lam, K. T., Samudra, G. S., & Liang, G., (2010). Device physics and characteristics of graphene nanoribbon tunneling FETs. *IEEE Transactions on Electron Devices, 57*(11), 3144–3152.

19. Das, D., & Rahaman, H., (2012). Modeling of IR-Drop induced delay fault in CNT and GNR power distribution networks. In: *2012 5th International Conference on Computers and Devices for Communication (CODEC)* (pp. 1–4). IEEE.

20. Dean, C. R., Young, A. F., Meric, I., Lee, C., Wang, L., Sorgenfrei, S., & Hone, J., (2010). Boron nitride substrates for high-quality graphene electronics. *Nature Nanotechnology, 5*(10), 722–726.

21. Dimitrakakis, G. K., Tylianakis, E., & Froudakis, G. E., (2008). Pillared graphene: A new 3-D network nanostructure for enhanced hydrogen storage. *Nano Letters, 8*(10), 3166–3170.

22. Dorgan, V. E., Bae, M. H., & Pop, E., (2010). Mobility and saturation velocity in graphene on SiO_2. *Applied Physics Letters, 97*(8), 082112.

23. Dresselhaus, G., Dresselhaus, M. S., & Saito, R., (1998). *Physical Properties of Carbon Nanotubes* (Vol. 1, pp. 1–33). World Scientific.

24. Du, X., Skachko, I., Barker, A., & Andrei, E. Y., (2008). Approaching ballistic transport in suspended graphene. *Nature Nanotechnology, 3*(8), 491–495.
25. Fang, T., Konar, A., Xing, H., & Jena, D., (2008). Mobility in semiconducting graphene nanoribbons: Phonon, impurity, and edge roughness scattering. *Physical Review B, 78*(20), 205403.
26. Farmer, D. B., Chiu, H. Y., Lin, Y. M., Jenkins, K. A., Xia, F., & Avouris, P., (2009). Utilization of a buffered dielectric to achieve high field-effect carrier mobility in graphene transistors. *Nano Letters, 9*(12), 4474–4478.
27. Farmer, D. B., Lin, Y. M., & Avouris, P., (2010). Graphene field-effect transistors with self-aligned gates. *Applied Physics Letters, 97*(1), 013103.
28. Fiori, G., & Iannaccone, G., (2007). Simulation of graphene nanoribbon field-effect transistors. *IEEE Electron Device Letters, 28*(8), 760–762.
29. Gannett, W., Regan, W., Watanabe, K., Taniguchi, T., Crommie, M. F., & Zettl, A., (2011). Boron nitride substrates for high mobility chemical vapor deposited graphene. *Applied Physics Letters, 98*(24), 242105.
30. Gargini, P., (2000). The international technology roadmap for semiconductors (ITRS): "Past, present and future." In: *GaAs IC Symposium. IEEE Gallium Arsenide Integrated Circuits Symposium. 22nd Annual Technical Digest 2000 (Cat. No. 00CH37084)* (pp. 3–5). IEEE.
31. Ghosh, D. S., Calizo, I., Teweldebrhan, D., Pokatilov, E. P., Nika, D. L., Balandin, A. A., & Lau, C. N., (2008). Extremely high thermal conductivity of graphene: Prospects for thermal management applications in nanoelectronic circuits. *Applied Physics Letters, 92*(15), 151911.
32. Ghosh, S., Bao, W., Nika, D. L., Subrina, S., Pokatilov, E. P., Lau, C. N., & Balandin, A. A., (2010). Dimensional crossover of thermal transport in few-layer graphene. *Nature Materials, 9*(7), 555–558.
33. Guerriero, E., Polloni, L., Rizzi, L. G., Bianchi, M., Mondello, G., & Sordan, R., (2012). Graphene audio voltage amplifier. *Small, 8*(3), 357–361.
34. Guo, J., Gunlycke, D., & White, C. T., (2008). Field-effect on spin-polarized transport in graphene nanoribbons. *Applied Physics Letters, 92*(16), 163109.

35. Gupta, G., Jalil, M. B. A., Yu, B., & Liang, G., (2012). Performance evaluation of electro-optic effect-based graphene transistors. *Nanoscale, 4*(20), 6365–6373.
36. Gupta, G., Zeng, M., Nurbawono, A., Huang, W., & Liang, G., (2016). *19 Applications of Graphene-Based Materials in Electronic Devices, 1*, 1–22.
37. Haji, N. S., Moravvej-Farshi, M. K., & Faez, R., (2012). Time-domain analysis of graphene nanoribbon interconnects based on transmission line model. *Iranian Journal of Electrical and Electronic Engineering, 8*(1), 37–44.
38. Han, M. Y., Özyilmaz, B., Zhang, Y., & Kim, P., (2007). Energy bandgap engineering of graphene nanoribbons. *Physical Review Letters, 98*(20), 206805.
39. Han, S. J., Garcia, A. V., Oida, S., Jenkins, K. A., & Haensch, W., (2014). Graphene radio frequency receiver integrated circuit. *Nature Communications, 5*(1), 1–6.
40. Hernandez, Y., Nicolosi, V., Lotya, M., Blighe, F. M., Sun, Z., De, S., & Boland, J. J., (2008). High-yield production of graphene by liquid-phase exfoliation of graphite. *Nature Nanotechnology, 3*(9), 563–568.
41. Huang, Y. C., Chang, C. P., & Lin, M. F., (2007). Magnetic and quantum confinement effects on electronic and optical properties of graphene ribbons. *Nanotechnology, 18*(49), 495401.
42. Iijima, S., (1991). Helical microtubules of graphitic carbon. *Nature, 354*(6348), 56–58.
43. Jain, N., Bansal, T., Durcan, C., & Yu, B., (2012). Effect of substrate on graphene-based interconnects. In: *2012 12th IEEE International Conference on Nanotechnology (IEEE-Nano)* (Vol. 1, pp. 1–4). IEEE.
44. Jiao, L., Wang, X., Diankov, G., Wang, H., & Dai, H., (2010). Facile synthesis of high-quality graphene nanoribbons. *Nature Nanotechnology, 5*(5), 321–325.
45. Jiao, L., Zhang, L., Wang, X., Diankov, G., & Dai, H., (2009). Narrow graphene nanoribbons from carbon nanotubes. *Nature, 458*(7240), 877–880.
46. Katsnelson, M. I., Novoselov, K. S., & Geim, A. K., (2006). Chiral tunnelling and the Klein paradox in graphene. *Nature Physics, 2*(9), 620–625.

47. Kim, W. Y., & Kim, K. S., (2008). Prediction of very large values of magnetoresistance in a graphene nanoribbon device. *Nature Nanotechnology, 3*(7), 408–412.
48. Koh, Y. K., Bae, M. H., Cahill, D. G., & Pop, E., (2010). Heat conduction across monolayer and few-layer graphenes. *Nano Letters, 10*(11), 4363–4368.
49. Kosynkin, D. V., Higginbotham, A. L., Sinitskii, A., Lomeda, J. R., Dimiev, A., Price, B. K., & Tour, J. M., (2009). Longitudinal unzipping of carbon nanotubes to form graphene nanoribbons. *Nature, 458*(7240), 872–876.
50. Kumar, S. B., Jalil, M. B. A., Tan, S. G., & Liang, G., (2010). Magnetoresistive effect in graphene nanoribbon due to magnetic field induced band gap modulation. *Journal of Applied Physics, 108*(3), 033709.
51. Kumar, S. B., Jalil, M. B. A., Tan, S. G., & Liang, G., (2010). The effect of magnetic field and disorders on the electronic transport in graphene nanoribbons. *Journal of Physics: Condensed Matter, 22*(37), 375303.
52. Kumar, V., Rakheja, S., & Naeemi, A., (2012). Performance and energy-per-bit modeling of multilayer graphene nanoribbon conductors. *IEEE Transactions on Electron Devices, 59*(10), 2753–2761.
53. Lam, K. T., Chin, S. K., Seah, D. W., Kumar, S. B., & Liang, G., (2010a). Effect of ribbon width and doping concentration on device performance of graphene nanoribbon tunneling field-effect transistors. *Japanese Journal of Applied Physics, 49*(4S), 04DJ10.
54. Lam, K. T., Seah, D., Chin, S. K., Kumar, S. B., Samudra, G., Yeo, Y. C., & Liang, G., (2010). A simulation study of graphene-nanoribbon tunneling FET with heterojunction channel. *IEEE Electron Device Letters, 31*(6), 555–557.
55. Lam, K. T., Sreenivas, V. P., & Liang, G., (2012). High frequency performance of graphene nanoribbon TFETs with phonon scattering. In: *15th International Workshop on Computational Electronics* (Vol. 1, pp. 221–222).
56. Leong, Z. Y., Lam, K. T., & Liang, G., (2009). Device performance of graphene nanoribbon field-effect transistors with edge roughness effects: A computational study. In: *2009 13th International Workshop on Computational Electronics* (pp. 1–4). IEEE.
57. Li, X., Wang, X., Zhang, L., Lee, S., & Dai, H., (2008). Chemically

derived, ultrasmooth graphene nanoribbon semiconductors. *Science, 319*(5867), 1229–1232.

58. Liang, G., Bala, K. S., Jalil, M. B. A., & Tan, S. G., (2011). High magnetoresistance at room temperature in pin graphene nanoribbons due to band-to-band tunneling effects. *Applied Physics Letters, 99*(8), 083107.
59. Liang, G., Khalid, S. B., & Lam, K. T., (2010). Influence of edge roughness on graphene nanoribbon resonant tunnelling diodes. *Journal of Physics D: Applied Physics, 43*(21), 215101.
60. Liang, G., Neophytou, N., Lundstrom, M. S., & Nikonov, D. E., (2007). Ballistic graphene nanoribbon metal-oxide-semiconductor field-effect transistors: A full real-space quantum transport simulation. *Journal of Applied Physics, 102*(5), 054307.
61. Liang, G., Neophytou, N., Lundstrom, M. S., & Nikonov, D. E., (2008). Contact effects in graphene nanoribbon transistors. *Nano Letters, 8*(7), 1819–1824.
62. Liang, G., Neophytou, N., Nikonov, D. E., & Lundstrom, M. S., (2007). Performance projections for ballistic graphene nanoribbon field-effect transistors. *IEEE Transactions on Electron Devices, 54*(4), 677–682.
63. Liang, Q., Yao, X., Wang, W., Liu, Y., & Wong, C. P., (2011). A three-dimensional vertically aligned functionalized multilayer graphene architecture: An approach for graphene-based thermal interfacial materials. *ACS Nano, 5*(3), 2392–2401.
64. Liao, L., & Duan, X., (2012). Graphene for radio frequency electronics. *Materials Today, 15*(7, 8), 328–338.
65. Liao, L., Bai, J., Cheng, R., Lin, Y. C., Jiang, S., Qu, Y., & Duan, X., (2010a). Sub-100 nm channel length graphene transistors. *Nano Letters, 10*(10), 3952–3956.
66. Liao, L., Bai, J., Cheng, R., Zhou, H., Liu, L., Liu, Y., & Duan, X., (2012). Scalable fabrication of self-aligned graphene transistors and circuits on glass. *Nano Letters, 12*(6), 2653–2657.
67. Liao, L., Lin, Y. C., Bao, M., Cheng, R., Bai, J., Liu, Y., & Duan, X., (2010b). High-speed graphene transistors with a self-aligned nanowire gate. *Nature, 467*(7313), 305–308.
68. Lin, Y. M., Jenkins, K. A., Valdes-Garcia, A., Small, J. P., Farmer, D. B., & Avouris, P., (2009). Operation of graphene transistors at gigahertz frequencies. *Nano Letters, 9*(1), 422–426.

69. Lin, Y. M., Valdes-Garcia, A., Han, S. J., Farmer, D. B., Meric, I., Sun, Y., & Jenkins, K. A., (2011). Wafer-scale graphene integrated circuit. *Science, 332*(6035), 1294–1297.
70. Low, T., & Appenzeller, J., (2009). Electronic transport properties of a tilted graphene p-n junction. *Physical Review B, 80*(15), 155406.
71. Lu, Y., & Guo, J., (2010). Local strain in tunneling transistors based on graphene nanoribbons. *Applied Physics Letters, 97*(7), 073105.
72. Luisier, M., & Klimeck, G., (2009). Performance analysis of statistical samples of graphene nanoribbon tunneling transistors with line edge roughness. *Applied Physics Letters, 94*(22), 223505.
73. Martin, J., Akerman, N., Ulbricht, G., Lohmann, T., Smet, J. V., Von, K. K., & Yacoby, A., (2008). Observation of electron-hole puddles in graphene using a scanning single-electron transistor. *Nature Physics, 4*(2), 144–148.
74. Mayorov, A. S., Gorbachev, R. V., Morozov, S. V., Britnell, L., Jalil, R., Ponomarenko, L. A., & Geim, A. K., (2011). Micrometer-scale ballistic transport in encapsulated graphene at room temperature. *Nano Letters, 11*(6), 2396–2399.
75. Meric, I., Dean, C. R., Petrone, N., Wang, L., Hone, J., Kim, P., & Shepard, K. L., (2013). Graphene field-effect transistors based on boron-nitride dielectrics. *Proceedings of the IEEE,101*(7), 1609–1619.
76. Naeemi, A., & Meindl, J. D., (2007). Conductance modeling for graphene nanoribbon (GNR) interconnects. *IEEE Electron Device Letters, 28*(5), 428–431.
77. Naeemi, A., & Meindl, J. D., (2008). Performance benchmarking for graphene nanoribbon, carbon nanotube, and Cu interconnects. In: *2008 International Interconnect Technology Conference* (Vol. 1, pp. 183–185). IEEE.
78. Naeemi, A., & Meindl, J. D., (2009). Compact physics-based circuit models for graphene nanoribbon interconnects. *IEEE Transactions on Electron Devices, 56*(9), 1822–1833.
79. Neto, A. C., Guinea, F., Peres, N. M., Novoselov, K. S., & Geim, A. K., (2009). The electronic properties of graphene. *Reviews of Modern Physics, 81*(1), 109.
80. Nika, D. L., Ghosh, S., Pokatilov, E. P., & Balandin, A. A., (2009). Lattice thermal conductivity of graphene flakes: Comparison with bulk graphite. *Applied Physics Letters, 94*(20), 203103.

81. Nirmalraj, P. N., Lutz, T., Kumar, S., Duesberg, G. S., & Boland, J. J., (2011). Nanoscale mapping of electrical resistivity and connectivity in graphene strips and networks. *Nano Letters, 11*(1), 16–22.
82. Nishad, A. K., & Sharma, R., (2013). Analytical time-domain models for performance optimization of multilayer GNR interconnects. *IEEE Journal of Selected Topics in Quantum Electronics, 20*(1), 17–24.
83. Novoselov, K. S., & Fal, V. I., (2012). In: Ko, L. Colombo, Gellert, P. R., Schwab, M. G., & Kim, K., (eds.), *Nature* (Vol. 490, No. 7419, pp. 192–200).
84. Novoselov, K. S., Geim, A. K., Morozov, S. V., Jiang, D., Katsnelson, M. I., Grigorieva, I., & Firsov, A. A., (2005). Two-dimensional gas of massless Dirac fermions in graphene. *Nature, 438*(7065), 197–200.
85. Novoselov, K. S., Geim, A. K., Morozov, S. V., Jiang, D., Zhang, Y., Dubonos, S. V., & Firsov, A. A., (2004). Electric field effect in atomically thin carbon films. *Science, 306*(5696), 666–669.
86. Obradovic, B., Kotlyar, R., Heinz, F., Matagne, P., Rakshit, T., Giles, M. D., & Nikonov, D. E., (2006). Analysis of graphene nanoribbons as a channel material for field-effect transistors. *Applied Physics Letters, 88*(14), 142102.
87. Ouyang, Y., Yoon, Y., & Guo, J., (2007). Scaling behaviors of graphene nanoribbon FETs: A three-dimensional quantum simulation study. *IEEE Transactions on Electron Devices, 54*(9), 2223–2231.
88. Rakheja, S., Kumar, V., & Naeemi, A., (2013). Evaluation of the potential performance of graphene nanoribbons as on-chip interconnects. *Proceedings of the IEEE,101*(7), 1740–1765.
89. Ramón, M. E., Parrish, K. N., Chowdhury, S. F., Magnuson, C. W., Movva, H. C., Ruoff, R. S., & Akinwande, D., (2012). Three-gigahertz graphene frequency doubler on quartz operating beyond the transit frequency. *IEEE Transactions on Nanotechnology, 11*(5), 877–883.
90. Sajjad, R. N., & Ghosh, A. W., (2011). High efficiency switching using graphene-based electron "optics." *Applied Physics Letters, 99*(12), 123101.
91. Sajjad, R. N., Polanco, C. A., & Ghosh, A. W., (2013). Atomistic deconstruction of current flow in graphene based hetero-junctions. *Journal of Computational Electronics, 12*(2), 232–247.
92. Santos, E. J., & Kaxiras, E., (2013). Electric-field dependence of the effective dielectric constant in graphene. *Nano Letters, 13*(3), 898–902.

93. Sarkar, D., Xu, C., Li, H., & Banerjee, K., (2011). High-frequency behavior of graphene-based interconnects—Part I: Impedance modeling. *IEEE Transactions on Electron Devices, 58*(3), 843–852.
94. Sarkar, D., Xu, C., Li, H., & Banerjee, K., (2011). High-frequency behavior of graphene-based interconnects—Part II: Impedance analysis and implications for inductor design. *IEEE Transactions on Electron Devices, 58*(3), 853–859.
95. Sarma, S. D., Adam, S., Hwang, E. H., & Rossi, E., (2011). Electronic transport in two-dimensional graphene. *Reviews of Modern Physics, 83*(2), 407.
96. Schwierz, F., (2010). Graphene transistors. *Nature Nanotechnology, 5*(7), 487.
97. Shishir, R. S., & Ferry, D. K., (2009). Velocity saturation in intrinsic graphene. *Journal of Physics: Condensed Matter, 21*(34), 344201.
98. Sohier, T., & Yu, B., (2011). Ultralow-voltage design of graphene p-n junction quantum reflective switch transistor. *Applied Physics Letters, 98*(21), 213104.
99. Son, Y. W., Cohen, M. L., & Louie, S. G., (2006). Energy gaps in graphene nanoribbons. *Physical Review Letters, 97*(21), 216803.
100. Son, Y. W., Cohen, M. L., & Louie, S. G., (2006). Half-metallic graphene nanoribbons. *Nature, 444*(7117), 347–349.
101. Soonmin, H., (2019). A review of nanostructured thin films for gas sensing and corrosion protection. *Mediterranean Journal of Chemistry, 7*(6), 433–451.
102. Sun, J. P., Haddad, G. I., Mazumder, P., & Schulman, J. N., (1998). Resonant tunneling diodes: Models and properties. *Proceedings of the IEEE,86*(4), 641–660.
103. Sutar, S., Comfort, E. S., Liu, J., Taniguchi, T., Watanabe, K., & Lee, J. U., (2012). Angle-dependent carrier transmission in graphene p-n junctions. *Nano Letters, 12*(9), 4460–4464.
104. Taur, Y., & Ning, T. H., (2013). *Fundamentals of Modern VLSI Devices* (Vol. 1, pp. 1–22). Cambridge University Press.
105. Teong, H., Lam, K. T., Khalid, S. B., & Liang, G., (2009). Shape effects in graphene nanoribbon resonant tunneling diodes: A computational study. *Journal of Applied Physics, 105*(8), 084317.
106. Tombros, N., Jozsa, C., Popinciuc, M., Jonkman, H. T., & Van, W. B. J., (2007). Electronic spin transport and spin precession in single

graphene layers at room temperature. *Nature, 448*(7153), 571–574.

107. Wakabayashi, K., (2001). Electronic transport properties of nano graphite ribbon junctions. *Physical Review B,64*(12), 125428.
108. Wallace, P. R., (1947). The band theory of graphite. *Physical Review, 71*(9), 622.
109. Wang, H., Hsu, A., Wu, J., Kong, J., & Palacios, T., (2010). Graphene-based ambipolar RF mixers. *IEEE Electron Device Letters, 31*(9), 906–908.
110. Wang, H., Nezich, D., Kong, J., & Palacios, T., (2009). Graphene frequency multipliers. *IEEE Electron Device Letters, 30*(5), 547–549.
111. Wang, X., Ouyang, Y., Li, X., Wang, H., Guo, J., & Dai, H., (2008). Room-temperature all-semiconducting sub-10-nm graphene nanoribbon field-effect transistors. *Physical Review Letters, 100*(20), 206803.
112. Wei, P., Bao, W., Pu, Y., Lau, C. N., & Shi, J., (2009). Anomalous thermoelectric transport of Dirac particles in graphene. *Physical Review Letters, 102*(16), 166808.
113. Williams, J. R., Low, T., Lundstrom, M. S., & Marcus, C. M., (2011). Gate-controlled guiding of electrons in graphene. *Nature Nanotechnology, 6*(4), 222–225.
114. Wu, Y., Jenkins, K. A., Valdes-Garcia, A., Farmer, D. B., Zhu, Y., Bol, A. A., & Lin, Y. M., (2012). State-of-the-art graphene high-frequency electronics. *Nano Letters, 12*(6), 3062–3067.
115. Wu, Y., Lin, Y. M., Bol, A. A., Jenkins, K. A., Xia, F., Farmer, D. B., & Avouris, P., (2011). High-frequency, scaled graphene transistors on diamond-like carbon. *Nature, 472*(7341), 74–78.
116. Wu, Z., (2011). Electronic fiber in graphene. *Applied Physics Letters, 98*(8), 082117.
117. Xia, F., Farmer, D. B., Lin, Y. M., & Avouris, P., (2010). Graphene field-effect transistors with high on/off current ratio and large transport bandgap at room temperature. *Nano Letters, 10*(2), 715–718.
118. Xu, C., Li, H., & Banerjee, K., (2009). Modeling, analysis, and design of graphene nanoribbon interconnects. *IEEE Transactions on Electron Devices, 56*(8), 1567–1578.
119. Yang, X., Liu, G., Balandin, A. A., & Mohanram, K., (2010). Triple-mode single-transistor graphene amplifier and its applications. *ACS Nano, 4*(10), 5532–5538.

120. Yoon, Y., & Guo, J., (2007). Effect of edge roughness in graphene nanoribbon transistors. *Applied Physics Letters, 91*(7), 073103.
121. Yoon, Y., & Salahuddin, S., (2010). Barrier-free tunneling in a carbon heterojunction transistor. *Applied Physics Letters, 97*(3), 033102.
122. Zeng, M., Feng, Y., & Liang, G., (2011b). Graphene-based spin caloritronics. *Nano Letters, 11*(3), 1369–1373.
123. Zeng, M., Shen, L., Su, H., Zhang, C., & Feng, Y., (2011a). Graphene-based spin logic gates. *Applied Physics Letters, 98*(9), 092110.
124. Zeng, M., Shen, L., Zhou, M., Zhang, C., & Feng, Y., (2011). Graphene-based bipolar spin diode and spin transistor: Rectification and amplification of spin-polarized current. *Physical Review B,83*(11), 115427.
125. Zhang, Y., Tan, Y. W., Stormer, H. L., & Kim, P., (2005). Experimental observation of the quantum hall effect and Berry's phase in graphene. *Nature, 438*(7065), 201–204.
126. Zhao, P., Chauhan, J., & Guo, J., (2009). Computational study of tunneling transistor based on graphene nanoribbon. *Nano Letters, 9*(2), 684–688.
127. Zheng, J., Wang, L., Quhe, R., Liu, Q., Li, H., Yu, D., & Lu, J., (2013). Sub-10 nm gate length graphene transistors: Operating at terahertz frequencies with current saturation. *Scientific Reports, 3*, 1314.
128. Zhu, W., Low, T., Farmer, D. B., Jenkins, K., Ek, B., & Avouris, P., (2013). Effect of dual gate control on the alternating current performance of graphene radiofrequency device. *Journal of Applied Physics, 114*(4), 044307.
129. Zhu, W., Yogeesh, M. N., Yang, S., Aldave, S. H., Kim, J. S., Sonde, S., & Akinwande, D., (2015). Flexible black phosphorus ambipolar transistors, circuits, and AM demodulator. *Nano Letters, 15*(3), 1883–1890.
130. Zuev, Y. M., Chang, W., & Kim, P., (2009). Thermoelectric and magneto thermoelectric transport measurements of graphene. *Physical Review Letters, 102*(9), 096807.

CHAPTER 4

Graphene-Based Electronic Sensors

CONTENTS

4.1. INTRODUCTION

Tracking down biological and chemical molecules having high selectivity and sensitivity is not only critical to the research field, but also for practical applications, like disease diagnosis, health care, and detection of gas leakage, etc. (Stewart et al., 2008; Ratinac et al., 2011). Many sensing techniques, like surface plasma resonance (SPR), surface-enhanced Raman spectroscopy (SERS), and electrochemistry, techniques are applied to obtain extraordinarily selective and sensitive, sensing devices having low costs used for targeting the tracking down of several harmful chemicals and especially, biomolecules in the aqueous environment. Electronic sensors working on field-effect transistors (FET) are favored because of their low cost, simple device configuration, high sensitivity, diminishment of devices, and real-time detection (Wang et al., 2011, 2012).

The basis of electronic detection is the conductivity change of FETs partially conducting networks on adsorption of the target molecules. Along with FETs, this plan of detection was exhibited by using bulk materials, like ion-sensitive polymer membranes gas-sensitive or metal oxides, networks. But, the bulk networks, that were applied in these earlier described planar devices, limited the interface among target molecules and network just on the network surface, resulting in low sensitivity of FETs sensors. So, their more applications are restricted (Willets and Van Duyne, 2007).

To raise the performance of FETs sensors, semiconductor nanomaterials are a perfect solution. Silicon nanowires (NWs) of single-crystal having one-dimensional are a distinctive example. The great switch properties (ON/OFF ratio greater than 10^7) of silicon NWs-based FETs are critical for the very high sensitivity of electronic sensors centered on this material. One more useful case is one-dimensional carbon nanotubes (CNTs). CNTs arrays, CNTs thin films, and specific semiconducting CNTs have been used in high-performance electronic sensors as networks. Furthermore, organic single crystals conjugated and polymer thin films are also being employed as networks in some further electronic sensors (Patolsky et al., 2006; Guo et al., 2010).

Graphene, the newly established two-dimensional (2D) material, because of its extensive range of electronics used in translucent electrodes is attracting cumulative attention, the active material in energy productionand packing, and network materials for FETs(Barsan and Weimar, 2003). Beyond, the greater electronic characteristics, graphene gives high biocompatibility and great surface area, flexibility, and superficial chemical operationalization related to CNTs, shaping it a perfect sensing stage. Most significantly, graphene's atomic depth shows that the total carbon atoms straight interrelate to the analyte, giving a decisive sensitivity, greater than that of CNTs and one-dimensional silicon NWs (Hu et al., 2010).

In this chapter, we shortly present the newly established, graphene-based biological, and chemical electronic sensors. The device working principles and structures of graphene-based FETs in the phases of gas and liquid are exhibited. The great detection of the bioactivities, biomolecules, and minor toxic gas molecules of living cells are debated. Furthermore, several kinds of graphene networks, such as reduced graphene oxide (rGO), graphene grownup via the CVD, and original graphene with their impacts on the device sensing actions and production are debated (Mcbride et al., 1978; Ramgir et al., 2010).

4.2. WORKING PRINCIPLE AND DEVICE CONFIGURATION

The common FET consists of a channel which is semiconducting among two metal electrodes in which one is the source and the other is the drain electrode, from where the current is introduced and gathered. By changing the potential of the gate, the channel's conductance can be controlled capacitively via a typically thin dielectric layer around 300 nm silicon dioxide (Kauffman and Star, 2008; He et al., 2012).

In a general p-type MOSFET (metal-oxide-semiconductor FET), to the gathering of holes the negative potential of gate leads (maximum charge carries), ensuring an improvement of the conductance of the channel, whereas to the depletion of holes the positive gate potential guides, and therefore a decline of the conductance. On the surface of the semiconducting channel in the electronic sensor, the molecule's adsorption either directly dopes the channel or alters its local surface potential resulting in the change of the conductance of the FET (Dai et al., 2002; Novoselov et al., 2004). This makes the FETs a useful sensing device having real-time capability, high sensitivity, and easily adjustable configuration. The gate electrode is detached in some situations, to make simpler the structure of the device, to make a chemo resistor. This configuration is suitable to produce graphene-based sensors for polymer substrates for flexible electronics uses. By the gate potential in spite of the deficiency of variation, the working principle of the chemo resistor is similar to a typical FETs sensor (Huang et al., 2011, 2012).

As exhibited in Figure 4.1(a), in a common gas sensing system, to the target gas, the network is directly exposed. The gas molecule's adsorption results as doping of semiconducting network, guiding to a change of conductance of the FETs device. The theoretical studies and experiments described that for the current response, the charge transmission from the adsorbed gas molecules towards the semiconducting channel is the leading mechanism, which is parallel to CNT-based gas sensors. Though some researches on CNT-based gas sensors recommended that the variation of metal-semiconductor contacts might also be the foremost mechanism, on graphene-based gas sensors, several pieces of research have ruled out this prospect (Hu et al., 2010; Yin et al., 2010a, 2010b).

In an aqueous environment for the detection of bio-species, the graphene-based FETs should work. As represented in Figure 4.1(b), the network of graphene is generally engrossed in a flow cell or sensing chamber, which is applied to restrict the solution. For the prevention of current leak from ionic conduction, the source and drain electrodes are isolated. Several insulators like silicone rubber, thin film of silicon dioxide, passivation of SU8, and silicone rubber/polydimethylsiloxane are applied in structures of several devices. In the solution, the gate electrode, typically AgCl/Ag or Pt, is placed. The potential of the gate is functional at the channel-solution boundary via the thin electric double layer capacitance made. By the ionic strength, the Debye length (or double-layer thickness) is determined in the solution, usually in 1 nm, which is even diluent as compared to the thin HfO_2 layer applied in the top-gate's high-performance graphene-based FETs. Usually, the solution-gate FETs are over two orders of scale more complex than the common back gate FETs (Wu et al., 2010, 2011).

For graphene-based biosensors, two main sensing mechanisms have been suggested in solution, like the doping effect and the electrostatic gating effect. The gating effect is the one in which on graphene the charged molecules adsorbed work as an extra capacitance of gating which modifies the graphene network's conductance. In contrast, the doping effect proposes a straight charge transmission among the graphene channel and adsorbed molecules, analogous to gas sensing (Bae et al., 2010).

4.3. CHOICE OF GRAPHENE MATERIALS

As displayed in Figure 4.2, in electronic sensors the three main types of graphene materials are used, like the original graphene synthesized by the mechanical exfoliation technique (shown in Figure 4.2(a)), graphenes grown by a chemical vapor deposition (CVD) method, stated to as CVD-graphene (shown in Figure 4.2(b)), and the rGO attained by the decrease of graphene oxide (GO), rising from the chemical exfoliation of graphite (Figure 4.2(c)).

In any solvent original graphene deprived of any functionalization is almost insoluble. Therefore, the "scotch tape" method is the only way to yield the original graphene channel which is through mechanical cleavage directly deposit graphene sheets onto the substrate (Sun et al., 2011). This type of graphene is not controllable in location, size, and shape. Therefore, the yield of devices is very less and such pristine graphene is more appropriate just for proof-of-concept and theoretical study demonstration (Pumera, 2011).

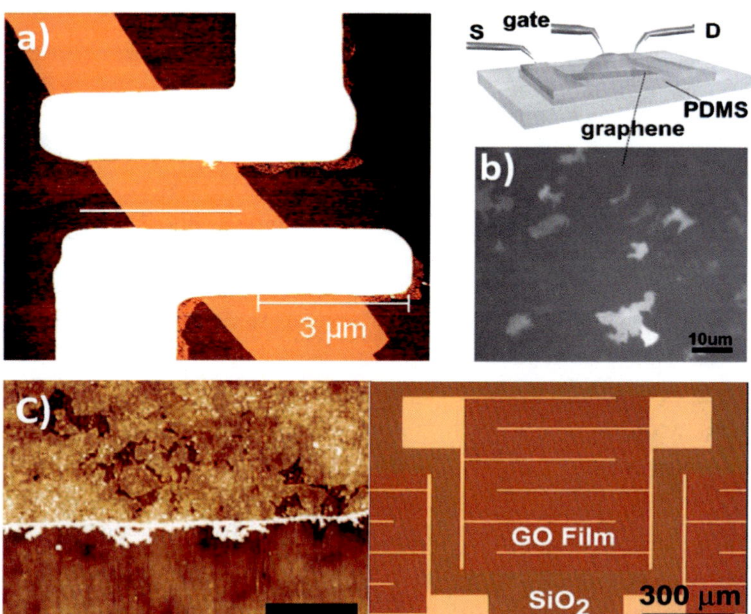

Figure 4.1. Graphene material types used in electronic sensors. (a) Image of a distinctive GFET centered on the Primeval graphene made by mechanical cleavage process having thickness 0.8 nm, by using AFM technique, related by Ti/Au electrodes and deposition is done using e-beam lithography. (b) CVD-Graphene films of chemical vapor deposited, and graphic diagram of the solution-gate FET images obtained by using optical microscopy. (c) rGO film image made by

a spin coating having a thickness of 6 nm, scale bar of 1 mm obtained by using AFM and image of rGO device which is electrically isolated and interdigitated Ti/Au contacts by using optical microscopy.

Source: https://pubs.acs.org/doi/10.1021/nl8013007.

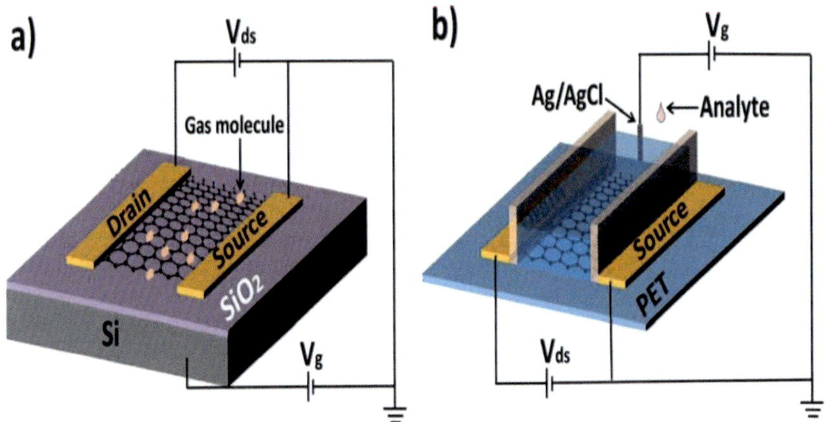

Figure 4.2. (a) A special back gate working as a gas sensor made of GFET on Si/SiO_2 substratum. (b) Special solution-gate of GFET on elastic polyethylene terephthalate substratum used for the biological and chemical sensor in aqueous solution.

Source: https://pubs.rsc.org/en/content/articlelanding/2012/sc/c2sc20205k.

In the CVD method, new advancement has been directed to the creation of a film of wafer-scale graphene on the metal substrates (like Cu, Ru, Ni) along with mono-layer produce as high as 95%. The high transparency and conductivity make the CVD-graphene appropriate for elastic electrodes applied in optoelectronic. On an important note, from the high-quality and high-yield CVD-graphene electronic sensors can also help, which through high reproducibility permits the huge production of devices. Nevertheless, meanwhile, the CVD-graphene can only be synthesized on specific substrates of metal, it needs extra transmission stages to form devices, guiding to destruction and contaminations of sheets of graphene. On the other hand, the epitaxial graphene grownup on silicon carbide (SiC) substrate, at a wafer size scale other kinds of thermally grown graphene, for the electronic sensing uses, has been applied. Nevertheless, device fabrication also requires further transfer steps (Burghard et al., 2009; Reddy et al., 2011).

The chemical exfoliation of graphite into GO at low cost founding a novel method to yield solution-processable graphene, which is vital for the bulk creation of electronics of graphene-based. A constant, homogeneous GO aqueous suspension is generally formed through an improved Hummers method. Thus, prepared GO is greatly oxidized and comprises alcohol groups, epoxy, and carboxylic acid. On GO, these functional groups by thermal or chemical reduction can be moderately detached to reinstate the conductivity and p-conjugated structure of GO in the formed rGO. While on rGO the residual functional groups will reduce the electronic performance by dropping its conductivity associated with the original graphene, the rGO-based electronic sensor may use from the improved chemical reaction and interaction among the analyte remaining and the functional groups (Yin et al., 2012; Schwierz, 2010).

4.4. GRAPHENE-BASED GAS SENSORS

Theoretic studies have assumed that the graphene channel's conduction can be changed by the gas molecule's adsorption, acting as acceptors or donors, on the surface of graphene. By experimentation, Schedin et al. (2007) invented a gas sensor that is depending on mechanically sliced mono-layer graphene. In their experiment, a four-probe measurement was performed to remove the effect of contact resistance. The final uncovering of a single NO_2 gas molecule was claimed, due to the great carrier movement of graphene and the very less noise of the device. In the meantime, Dan et al.claimed that the inherent response of graphene to the gas adsorption was rather slight, and from the chemical contamination, the main device response came, on graphene which doped the graphene channel as well as improved the adsorption of the molecules of gas (Leenaerts et al., 2008; Sudibya et al., 2011).

On graphene-based FETs, the continuation work was mainly on the scalable graphene-based, such as rGO, epitaxial graphene, and CVD-graphene. For instance, one new research presented that 3D graphene foam along with some layers' graphene grown via CVD technique is capable to spot NO_2 gas to an order of magnitude lesser concentration related to the viable polypyrrole sensor (Figure 4.3(a)). It is important to write that with three-dimensional (3D) graphene foam the device production is scalably deprived of the need for graphene transfer steps (Cao et al., 2011; Yavari et al., 2011). In the meantime, Chen et al. (2011) described by applying plasma-enhanced CVD, the through growth of vertically oriented graphene sheets on the silver electrodes to produce graphene-based FETs sensors for the detection of NH_3 and NO_2. Furthermore, Nomani et al. (2010) directly made up a NO_2 sensor by applying the epitaxial graphene on the semi-insulating 6H-SiC surface deprived of any transfer phase (Chen et al., 2011; Yu et al., 2011).

Though gas sensors centered on distinctly rGO sheets have been displayed, the main benefit of rGO in its solution processibility occurs, which is well-matched with several thin-film technologies. Fowler et al. (2008) made a chemo resistor sensor by the use of rGO sheets spin-coated on the interdigitated electrode array. The rGO sheets are generally monolayer and disjointed, linked through the electrode array. The various current responses to NH_3 and NO_2 maintained the transfer of charge mechanism. As the graphene-based FETs have functioned at the p-type region, the electron-removing NO_2 would create a reduction of resistance, whereas the electron-donating NH_3 would create a rise of resistance. Furthermore, this chemo resistor was capable to spot DNT (2,4-dinitrotoluene), a comparatively volatile component originates in the TNT (trinitrotoluene) explosive, at the ppb stage (Lu et al., 2009; Dua et al., 2010).

Other various production approaches were applied to produce gas sensors centered on rGO thin films. Robinson et al. (2008) produced the graphene-based FETs via direct deposition of Au electrodes by spin-coating on the 2–4 nm rGO, thin films obtained. They came to know that the rGO-based device shown meaningly lesser noise than did the device depending on CNT links, ensuing in dinitro toluene's a very low detection limit. Their research also exposed the various gas adsorption rates on the residual sp^2-carbon backbone of rGO and the oxygen-containing groups (Johnson et al., 2010; Li et al., 2011).

Via the viable inkjet-printing method the solution processibility of rGO makes it probable to produced rGO-based sensors, where the usage ink was the surfactant-protected rGO aqueous suspension attained via a green reduction of GO through vitamin C. On the PET (polyethylene terephthalate) film, the reduced oxide channels were decorated through inkjet printing. Sip socket leads were applied to join the rGO networks to make a lightweight, flexible sensor array (Wu et al., 2010). A planned detection threshold of 400 ppt for NO_2 based on the signal-to-noise ratio of three was displayed. It is important to mention that while rGO the film used in this research was thick (greater than 100 nm) as compared to other rGO sensors, it did not compromise the gas sensor's sensitivity.

In recent times, the current dielectrophoresis technique, formerly applied to produce CNT-based electronics, has been applied to make rGO channels among source and drain electrodes to form graphene-based field-effect sensors (Wu et al., 2010). The Au electrodes were protected with CVD-graphene earlier the deposition of rGO film among them. Among the rGO network and metal electrodes, the graphene coverage dropped the contact resistance. The rGO sheets were functionalized through Pd nanoparticles (PdNPs) to increase gas sensitivity (Chu et al., 2011).

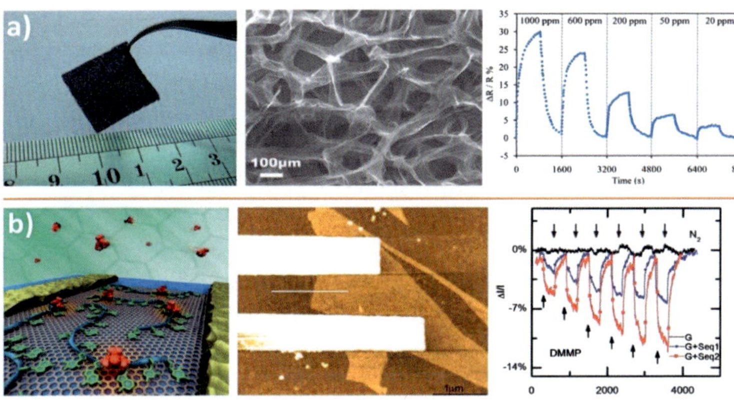

Figure 4.3. (a) SEM images of three-dimensional graphene foam full-grown using chemical vapor deposition and the real-time detection of NH_3 at several concentrations. (b) Graphic explanation of adsorption of single-stranded DNA on primeval graphene as a network for gas sensing, AFM image of adsorption of single-stranded DNA on primeval graphene among gold electrodes, and real-time detection of ssDNA-deprived dimethyl methyl phosphonate and two distinct sequences of ssDNA.

Source: https://repository.upenn.edu/cgi/viewcontent.cgi?article=1009&context=physics_papers.

Also, the usually displayed NH_3 and NO_x gas sensors, graphene-based field-effect-based sensors were used in the detection of H_2S, alcohol, H_2O, and H_2. In spite of the great sensitivity of gas sensors of graphene, in making practical graphene-based field-effect sensors challenge yet persist. For instance, their time of recovery is comparatively long, starting from minutes to hours. Lu et al. (2010) by joining the graphene network along with ssDNA (single-stranded DNA) tried to resolve this challenge as a sensitizing agent. As displayed in Figure 4.3(b), the non-covalent functionalization of ssDNA works as a concentrator of analyte molecules and water on the hydrophobic surface and chemically inert of original graphene, ensuing in a quick response along with complete and fast recovery after sensing. In the meantime, another research stated that the rGO-based FETs exhibited very fast recovery and response when at n-type region it was operated, rather than the p-type region normally used (Zhang et al., 2010; Yi et al., 2011). Device-to device variation is another common disadvantage of graphene-based field-effect sensors. To solve this challenge, thin-film devices based on homogenous graphene, high-quality, and large-area films are preferred. To conclude, a more thorough mechanistic study is still essential to further comprehend graphene-based field-effect sensors, to improve their reliability and performance in practical gas sensing uses (Cuong et al., 2010).

4.5. GRAPHENE-BASED PH SENSORS AND BIOSENSORS

The excellent chemical stability, good biocompatibility, and large surface area in an aqueous environment make graphene and its byproducts perfect platforms for biosensors (Lu et al., 2011). Furthermore, the facile noncovalent or covalent surface functionalization with different receptors keeps them easy to conjugate for precise bio detections (Lu et al., 2011).

In an aqueous environment, the graphene-based FET was effectively applied as a pH sensor. In this chapter, epitaxially grownup triple-, double-, single-layer graphene sheets were applied in solution-gate FETs along with AgCl/Ag as a gate electrode. A positive change of the transfer curve (solution-gate potential vs. drain to source current.) was created when the pH value of the raised buffer solution improved from 2 to 12. Triple-, double-, and single-layer graphene-based FETs at 99 mV pH exhibited the same pH sensitivity. The sensing mechanism rises from the change of ions (H_3O and OH) at different pH values adsorbed on the graphene surface, which at the graphene-solution boundary altered the electrochemical double layer (Liu et al., 2012).

At a graphene-based FET remarkably, a slighter pH sensitivity of 20 mV pH was detected depending on mechanically exfoliated monolayer graphene appending on two gold electrodes. In electronic sensing, the hanging graphene network was intended to decrease the noise. Nevertheless, no insulation was applied in experiments, which may bring ionic leak on the simple Au electrodes. Furthermore, the same change of transmission curves upon the change of pH value was detected at another type of graphene-based FET sensor, where silicone rubber was applied as the insulator among the graphene network and the Au electrode. So, the effect of an ionic leak from the barrier solution to electrodes was neglectable. Though, the swing of the transfer curve was found to be uneven from device to device, creation the pH measurement undependable. The purpose might rise from the charged impurities adsorbed on the top of the graphene surface and the boundary underneath the substrate and graphene (Fu et al., 2011).

The selective growth of graphite electrodes and monolithically combined graphene networks was realized through a CVD method on a spatially decorated heterogeneous metal catalyst. The graphene-based FET array can be transported on several flexible substrates to make a multiplexed sensor array. An immediate and real-time pH finding through 10 devices on a solo sensor array was displayed with brilliant reproducibility. As graphene has ambipolar characteristics, the sensor can be functioned at both p-type regions and n-type by just varying the solution-gate potential (Ristein et al., 2010).

Nevertheless, recently, Fu et al.(2011) claimed that to the pH value graphene-based FETs were insensitive. They argued that no precise compulsory of any ions was estimated in a fresh graphene surface. Depending on their work, the extremely sensitive pH detection stated overhead is most probably because of the graphene sheet's defects. These imperfections, generally the oxygen-containing groups, for protons, work as required cites, ensuing in the pH response. Through the thoughtful overview of defects, this was more confirmed in graphene through mild UV-ozone action, which enlarged the sensor's pH sensitivity. So, the alteration in pH sensitivity (12–99 mV pH^1) from the beforehand stated pH sensors just replicates the change in the value of the graphene network. In other research, in the graphene network flaws were found to be induced when the solution-gate potential was more than +0.3 V, which may negotiate the dependability of the graphene-based FET-based pH sensors (Cote et al., 2009).

In an aqueous environment, the effective operation of graphene-based FET sensors have improved biosensing researches. For instance, Li et al. applied it as a DNA sensor. In this chapter, with Au nanoparticle's millimeter-sized CVD graphene, was initially functionalized. The thiolated probe ssDNA was then stopped on AuNPs via Au-thiol bonding. By using the back-gate graphene-based FET, the uncovering of corresponding target ssDNA was understood by checking the change of the transfer curve after and before DNA hybridizing. Like the gas sensor, the sensing process was supposed to be the straight n-doping outcome of the adsorbed DNA to the graphene network. This was verified through the Dirac point and the negative shift of the transfer curve (Yin et al., 2012).

Though, the above-mentioned DNA sensor is not real-time as the graphene-based FET is functioned in back-gate mode to check the change of transfer curves. In this work, the Langmuir-Blodgett (LB) technique was applied for the scalable production of high-quality rGO dense films with one-to-two-layer thickness (Huang et al., 2010). Along with PtNPs, the rGO network was functionalized through in situ photochemical decreases, to stop the thiolated probe ssDNA via the Pt-thiol bonding. Through the real-time monitoring of current change, the exposure was understood upon the accumulation of the target ssDNA. Displaying no response to the uneven ssDNA, an intended detection threshold of 2.5 nM for the corresponding target ssDNA was attained. In several methods to real-time DNA detection, a few-layer GO film was spin-coated on decorated thin gold (10 nm) electrodes to organize a graphene-based FET sensing chip. Before the chemical decrease of graphene, oxide ethylenediamine was applied to shield the active required, sites, and give the main amine groups after the decrease. These prime amine groups were applied to stop the aminated probe ssDNA via the glutaraldehyde used as the linker (Mao et al., 2017).

Besides DNA, graphene-based FET sensors have been applied for the real-time exposure of other biomolecules like proteins and glucose. Ohno et al. (2010) exhibited that their pH sensor is suitable to distinguish in physiological barrier solution the adsorption of BSA (bovine serum albumin). A real-time sub-nM uncovering of BSA was exhibited. To the electrostatic gating effect, the sensing procedure was ascribed, according to which the negatively charged proteins gathered on the surface of graphene which acted as further negative gating, which when it was functioned by p-type region improved the conductance of the graphene-based FET. It is very important that the graphene-based FET shared the use of the solution gate and back gate to study the several gating effects.

The above-mentioned protein sensor is constructed on the non-specific adsorption of proteins on the surface of graphene. Though, the operationalization of graphene with receptors is essential to precisely spot target protein. For instance, Ohno et al. (2010) operationalized graphene-based FET along with IgE (immune globulin E) aptamers to attain a careful detection of IgE proteins. The operationalization of aptamers was comprehended by attaching 1-pyrenebutanoic acid succinimidyl ester (1-pbase), a linker molecule frequently applied to operationalize CNTs, on the surface of graphene through p-p interface. To block and deactivate the residual extra reactive groups, ethanolamine was then applied on the surface of graphene. For example, the thermal development of graphite graphene sheets was attained functionalized with precise antibodies to specifically detect the PSA (prostate-specific antigen), cancer biomarker (Mao et al., 2010). To avoid non-specific adsorption a similar linker molecule (1-pbase) was applied for the functionalization, whereas BSA passivation was applied (Mao et al., 2010, 2017).

The chemical functionalization of decorative reduces graphene oxide thin film along with biotin to precisely notice avidin, the rGO thin film has been initially coated by PEG (polyethylene glycol) and PEI (poly ethylene imine). The PEI has been applied to stop the receptor (biotin) via the amine groups on its backbone of the polymer. PEG was then used to avoid the non-specific adsorption of further proteins. The rGO-based sensor was capable of distinguishing avidin with high selectivity and sensitivity in the barrier solution. Furthermore, the sensor contains a full rGO formation, which shows that the rGO film was applied as both the semiconducting network and the electrodes. The exceptional elasticity of reducing graphene oxide made the entire device completely flexible and even for 5000 bending cycles, the device worked well. Similarly, the solution processibility of rGO made the production very reliable and very cheap. One more exhibition of the rGO-based protein sensor was described by Mao et al. (2010). As displayed in Figure 4.4(a), a thermally rGO film was operationalized along with the AuNP antibody (anti-IgG) conjugate solution to precisely distinguish the target protein (Mao et al., 2010). An obstructive buffer comprising BSA and fish gelatin was applied to avoid non-specific adsorption. Through a distinctive back-gate, FETs test the detection was attained. Nevertheless, upon adsorption of target proteins, a reduction of the drain to source current (I_{ds}) was detected (Figure 4.4(a)), in its place of the generally practical shift of the transfer curve in the DNA and pH sensors.

Related to the biomolecule's detection, the discovery of living cells is very interesting as the interface among the living cell membranes and graphene network is very complex. Overcurrent years, CNTs and one-dimensional single-crystal silicon NWsin the observation of living cells and their bioactivities have been widely studied(Li et al., 2010, 2012). However, graphene gives an improved chance to study the cell-nanomaterial boundary because it has a 2D structure offers a homogeneous interaction with the 2D cell membrane.

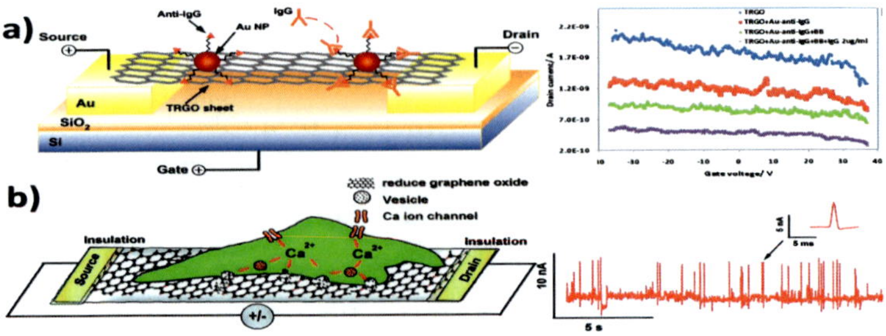

Figure 4.4. (a) Reduced graphene oxide field-effect transistors-based protein sensor schematic design. Anti-IgG is attached over the reduced graphene oxide surface over AuNPs and acts as a precise recognition group for IgG compulsory. The detection was understood by measuring the Ids-Vg curve after and before the adsorption of the target protein. Modified with authorization from ref. 88. (b) Graphic illustration of the boundary among a reduced graphene oxide field-effect transistor and PC12 cell, and the active emission of catecholamines VIA PC12 cell activated by potassium ions. The detection was apprehended by the real-time monitoring of Ids throughout the dynamic excretion of catecholamines.

Source: https://pubs.rsc.org/en/content/articlelanding/2011/cc/c1cc12648b.

For instance, Cohen-Karni et al. (2010) described a graphene-based FET sensor in order to detect bioelectrical signals through living chicken cells. On the silicon dioxide substrate, a back-gate graphene-based FET was invented by using original graphene by e-beam lithography. A single crystal silicon NWs FET was fabricated along with the graphene-based FET in order to equivalence their sensing performances. The graphene-based FETs sensor was from the impulsive beating of embryonic chicken cardiomyocyte capable to detect the distinct extracellular signals. It displayed a much-improved fixed signal than the usual planar device, nevertheless a lesser 3D resolution than the single-crystal silicon NWs FET because of its huge contact area. In one more example, Ang et al. (2008) applied a decorated carbon vapor deposited-graphene fabricated through photolithography in a PDMS flow cell to organize a graphene-based FET sensor array. After the operationalization of the surface of graphene along with CD36 receptor proteins, the sensor array was capable of distinguishing the release and capture at the single-cell level of malaria-infected red blood cells. The same work has been described by Hess et al. (2013) in which through the transfer of photolithography-patterned chemical vapor deposited graphene a sensor array was made-up on a navy substrate pre-coated along with a thermally dispersed Au/Ti electrode array. The array of the sensor along with 16 distinct graphene-based FETs was capable to trail the potential achievement of a cardiomyocyte-such as HL-1 cell crossways the sensor array, display a good-resolved signal (He et al., 2012).

The research applied a rGO-based solution-gate FET to examine living cell actions on rGO. As displayed in Figure 4.4(b), the devices were invented lied on large-scale micro decorated rGO tinny films along with a thickness of 1–3 nm. To acquire a friendly contact living neuron cells were greatly cultivated on the rGO network. The concentrated GOFET was capable to detect the adsorption of hormonal catecholamine molecules and those concealed from the living neuron cells having a great signal-to-noise ratio. Furthermore, the rGO FET could be produced on the elastic PET substrate and during bending work well, which might be beneficial in complex in vivo biosensing (Mohanty and Berry, 2008).

In recent times, Novoselov et al. (2005)described an excellent detection of cell bioactivity centered on the cell-induced mechanical distortion of the rGO network. In this research, the yeast cell was involved in the rGO network via the chemical operationalization. An alcohol-induced reduction of the yeast cells brings about the distortion of the rGO sheet, guiding to a conductance decline of the rGO network (Gong et al., 2010; Huang et al., 2011).

Furthermore, a GOFET centered on amine-modified rGO has been applied to distinguish the addon of bacteria. The add-on of a single bacterium brought the 42% rise of the conductance because of the p-doping effect. This great sensitivity can be described through the good interaction among the negatively charged bacterial cell wall and the minimized rGO surface. In one more instance, the CVD-graphene network operationalized along with antibodies of E. coli permitted a precise finding of E. coli bacteria in the existence of one more bacterium (Nolan and Lippard, 2007; Zhang et al., 2010).

4.6. GRAPHENE-BASED HEAVY-METAL SENSING

In addition to bio-species in an aqueous environment, the ability to work the graphene-based FET sensors are also capable of the solution for the detection of metal ions. Along with a graphene-based FET a Hg^{2+} sensor has been displayed based on the primeval graphene network functionalized from solution with alkyl thiol to detention Hg^{2+} ions (Cho et al., 2011). Because this device was worked in the back gate along with ex-situ reflection of the change of transfer curve, it is not a real-time detection. Furthermore, depended on the thiol-metal bonding this detection is general because it could be restricted by the existence of further metal ions (Radisavljevic et al., 2011).

Lately, the researcher presented by applying chemically functionalized rGO micropatterns the particular detection of heavy metal ions. The decorated rGO network was operationalized along with metallothionein type II via 1-pbase worked as the linker molecule. Metallothionein type II is a protein that can precisely fix to substantial metal ions along with great selectivity and affinity (He et al., 2012). This sensor was capable of distinguishing Hg^{2+} at stages lower than 1 nM (0.2 ppb) deprived of retort to extra metal ions like K^{+}, $Ca^{2+,}$ or Mg^{2+}. This sensitivity is greater than that of many fluorescence-based techniques and similar to the lately printed stripping voltammetry technique(Coleman et al., 2011). By swapping the operationalization as of metallothionein type II towards calmodulin, a Ca^{2+} compulsory protein, the sensor was capable to distinctively distinguish Ca^{2+} between other ions. Rather than the straight gating impact of adsorbed metal ions, the recent modification rose from conformational modification of the operationalized protein upon metal required. These conformational variations carried the negatively charged protein nearer to the network of graphene, thus firming its gating effect (Coleman et al., 2011; Zeng et al., 2011).

REFERENCES

1. Agarwal, S., Zhou, X., Ye, F., He, Q., Chen, G. C., Soo, J., & Chen, P., (2010). Interfacing live cells with nanocarbon substrates. *Langmuir, 26*(4), 2244–2247.
2. Ang, P. K., Chen, W., Wee, A. T. S., & Loh, K. P., (2008). Solution-gated epitaxial graphene as pH sensor. *Journal of the American Chemical Society, 130*(44), 14392–14393.
3. Ang, P. K., Li, A., Jaiswal, M., Wang, Y., Hou, H. W., Thong, J. T., & Loh, K. P., (2011). Flow sensing of single-cell by graphene transistor in a microfluidic channel. *Nano Letters, 11*(12), 5240–5246.
4. Bae, S., Kim, H., Lee, Y., Xu, X., Park, J. S., Zheng, Y., & Kim, Y. J., (2010). Roll-to-roll production of 30-inch graphene films for transparent electrodes. *Nature Nanotechnology, 5*(8), 574.
5. Barsan, N., & Weimar, U., (2003). Understanding the fundamental principles of metal oxide-based gas sensors; the example of CO sensing with SnO_2 sensors in the presence of humidity. *Journal of Physics: Condensed Matter, 15*(20), R813.
6. Burghard, M., Klauk, H., & Kern, K., (2009). Carbon□based field□effect transistors for nanoelectronics. *Advanced Materials, 21*(25, 26), 2586–2600.
7. Cao, X., He, Q., Shi, W., Li, B., Zeng, Z., Shi, Y., & Zhang, H., (2011). Graphene oxide as a carbon source for controlled growth of carbon nanowires. *Small, 7*(9), 1199–1202.
8. Cao, X., Shi, Y., Shi, W., Lu, G., Huang, X., Yan, Q., & Zhang, H., (2011). Preparation of novel 3D graphene networks for supercapacitor applications. *Small, 7*(22), 3163–3168.
9. Chakraborty, T., Manaselyan, A., Barseghyan, M., & Laroze, D., (2018). Controllable continuous evolution of electronic states in a single quantum ring. *Physical Review B,97*(4), 041304.
10. Chang, H., Lee, J. D., Lee, S. M., & Lee, Y. H., (2001). Adsorption of NH_3 and NO_2 molecules on carbon nanotubes. *Applied Physics Letters, 79*(23), 3863–3865.
11. Chen, Z., Ren, W., Gao, L., Liu, B., Pei, S., & Cheng, H. M., (2011). Three-dimensional flexible and conductive interconnected graphene networks grown by chemical vapor deposition. *Nature Materials, 10*(6), 424–428.

12. Cheng, Z., Li, Q., Li, Z., Zhou, Q., & Fang, Y., (2010). Suspended graphene sensors with improved signal and reduced noise. *Nano Letters, 10*(5), 1864–1868.
13. Cho, S., Butch, N. P., Paglione, J., & Fuhrer, M. S., (2011). Insulating behavior in ultrathin bismuth selenide field-effect transistors. *Nano Letters, 11*(5), 1925–1927.
14. Chu, B. H., Lo, C. F., Nicolosi, J., Chang, C. Y., Chen, V., Strupinski, W., & Ren, F., (2011). Hydrogen detection using platinum-coated graphene grown on SiC. *Sensors and Actuators B: Chemical, 157*(2), 500–503.
15. Cohen-Karni, T., Qing, Q., Li, Q., Fang, Y., & Lieber, C. M., (2010). Graphene and nanowire transistors for cellular interfaces and electrical recording. *Nano Letters, 10*(3), 1098–1102.
16. Coleman, J. N., Lotya, M., O'Neill, A., Bergin, S. D., King, P. J., Khan, U., & Shvets, I. V., (2011). Two-dimensional nanosheets produced by liquid exfoliation of layered materials. *Science, 331*(6017), 568–571.
17. Cote, L. J., Kim, F., & Huang, J., (2009). Langmuir-Blodgett assembly of graphite oxide single layers. *Journal of the American Chemical Society, 131*(3), 1043–1049.
18. Cuong, T. V., Pham, V. H., Chung, J. S., Shin, E. W., Yoo, D. H., Hahn, S. H., & Kohl, P. A., (2010). Solution-processed ZnO-chemically converted graphene gas sensor. *Materials Letters, 64*(22), 2479–2482.
19. Dai, L., Soundarrajan, P., & Kim, T., (2002). Sensors and sensor arrays based on conjugated polymers and carbon nanotubes. *Pure and Applied Chemistry, 74*(9), 1753–1772.
20. Dan, Y., Lu, Y., Kybert, N. J., Luo, Z., & Johnson, A. C., (2009). Intrinsic response of graphene vapor sensors. *Nano Letters, 9*(4), 1472–1475.
21. Dong, X., Shi, Y., Huang, W., Chen, P., & Li, L. J., (2010). Electrical detection of DNA hybridization with single□base specificity using transistors based on CVD□grown graphene sheets. *Advanced Materials, 22*(14), 1649–1653.
22. Dua, V., Surwade, S. P., Ammu, S., Agnihotra, S. R., Jain, S., Roberts, K. E., & Manohar, S. K., (2010). All□organic vapor sensor using inkjet□printed reduced graphene oxide. *Angewandte Chemie International Edition, 49*(12), 2154–2157.

23. Fowler, J. D., Allen, M. J., Tung, V. C., Yang, Y., Kaner, R. B., & Weiller, B. H., (2009). Practical chemical sensors from chemically derived graphene. *ACS Nano, 3*(2), 301–306.

24. Fu, W., Nef, C., Knopfmacher, O., Tarasov, A., Weiss, M., Calame, M., & Schönenberger, C., (2011). Graphene transistors are insensitive to pH changes in solution. *Nano Letters, 11*(9), 3597–3600.

25. Gong, J., Zhou, T., Song, D., & Zhang, L., (2010). Monodispersed Au nanoparticles decorated graphene as an enhanced sensing platform for ultrasensitive stripping voltammetric detection of mercury (II). *Sensors and actuators B: Chemical, 150*(2), 491–497.

26. Guo, Y., Yu, G., & Liu, Y., (2010). Functional organic field□effect transistors. *Advanced Materials, 22*(40), 4427–4447.

27. He, Q., Wu, S., Gao, S., Cao, X., Yin, Z., Li, H., & Zhang, H., (2011). Transparent, flexible, all-reduced graphene oxide thin-film transistors. *ACS Nano, 5*(6), 5038–5044.

28. He, Q., Wu, S., Yin, Z., & Zhang, H., (2012). Graphene-based electronic sensors. *Chemical Science, 3*(6), 1764–1772.

29. Hess, L. H., Seifert, M., & Garrido, J. A., (2013). Graphene transistors for bioelectronics. *Proceedings of the IEEE,101*(7), 1780–1792.

30. Hu, L., Kim, H. S., Lee, J. Y., Peumans, P., & Cui, Y., (2010). Scalable coating and properties of transparent, flexible, silver nanowire electrodes. *ACS Nano, 4*(5), 2955–2963.

31. Hu, P., Zhang, J., Li, L. E., Wang, Z., O'Neill, W., & Estrela, P., (2010). Carbon nanostructure-based field-effect transistors for label-free chemical/biological sensors. *Sensors, 10*(5), 5133–5159.

32. Huang, X., Qi, X., Boey, F., & Zhang, H., (2012). Graphene-based composites. *Chemical Society Reviews, 41*(2), 666–686.

33. Huang, X., Yin, Z., Wu, S., Qi, X., He, Q., Zhang, Q., & Zhang, H., (2011). Graphene□based materials: Synthesis, characterization, properties, and applications. *Small, 7*(14), 1876–1902.

34. Huang, Y., Dong, X., Liu, Y., Li, L. J., & Chen, P., (2011). Graphene-based biosensors for detection of bacteria and their metabolic activities. *Journal of Materials Chemistry, 21*(33), 12358–12362.

35. Huang, Y., Dong, X., Shi, Y., Li, C. M., Li, L. J., & Chen, P., (2010). Nanoelectronic biosensors based on CVD grown graphene. *Nanoscale, 2*(8), 1485–1488.

36. Johnson, J. L., Behnam, A., Pearton, S. J., & Ural, A., (2010). Hydrogen sensing using Pd□functionalized multi□layer graphene nanoribbon networks. *Advanced Materials, 22*(43), 4877–4880.
37. Katz, H. E., (2004). Chemically sensitive field□effect transistors and chemiresistors: New materials and device structures. *Electroanalysis: An International Journal Devoted to Fundamental and Practical Aspects of Electroanalysis, 16*(22), 1837–1842.
38. Kauffman, D. R., & Star, A., (2008). Electronically monitoring biological interactions with carbon nanotube field-effect transistors. *Chemical Society Reviews, 37*(6), 1197–1206.
39. Lee, M. S., Lee, K., Kim, S. Y., Lee, H., Park, J., Choi, K. H., & Park, J. U., (2013). High-performance, transparent, and stretchable electrodes using graphene-metal nanowire hybrid structures. *Nano Letters, 13*(6), 2814–2821.
40. Leenaerts, O., Partoens, B., & Peeters, F. M., (2008). Paramagnetic adsorbates on graphene: A charge transfer analysis. *Applied Physics Letters, 92*(24), 243125.
41. Li, D., Müller, M. B., Gilje, S., Kaner, R. B., & Wallace, G. G., (2008). Processable aqueous dispersions of graphene nanosheets. *Nature Nanotechnology, 3*(2), 101–105.
42. Li, H., Lu, G., Yin, Z., He, Q., Li, H., Zhang, Q., & Zhang, H., (2012). Optical identification of single□and few□layer MoS_2 sheets. *Small, 8*(5), 682–686.
43. Li, R., Fan, W., Tian, G., Zhu, H., He, L., Cai, J., & Zhang, Z., (2010). The sequence and de novo assembly of the giant panda genome. *Nature, 463*(7279), 311–317.
44. Li, W., Geng, X., Guo, Y., Rong, J., Gong, Y., Wu, L., & Sun, M., (2011). Reduced graphene oxide electrically contacted graphene sensor for highly sensitive nitric oxide detection. *ACS Nano, 5*(9), 6955–6961.
45. Li, X., Cai, W., An, J., Kim, S., Nah, J., Yang, D., & Banerjee, S. K., (2009). Large-area synthesis of high-quality and uniform graphene films on copper foils. *Science, 324*(5932), 1312–1314.
46. Liang, X., Fu, Z., & Chou, S. Y., (2007). Graphene transistors fabricated via transfer-printing in device active-areas on large wafer. *Nano Letters, 7*(12), 3840–3844.

47. Liu, Y., Dong, X., & Chen, P., (2012). Biological and chemical sensors based on graphene materials. *Chemical Society Reviews, 41*(6), 2283–2307.

48. Lu, G., Ocola, L. E., & Chen, J., (2009). Gas detection using low-temperature reduced graphene oxide sheets. *Applied Physics Letters, 94*(8), 083111.

49. Lu, G., Park, S., Yu, K., Ruoff, R. S., Ocola, L. E., Rosenmann, D., & Chen, J., (2011). Toward practical gas sensing with highly reduced graphene oxide: A new signal processing method to circumvent run-to-run and device-to-device variations. *ACS Nano, 5*(2), 1154–1164.

50. Lu, G., Yu, K., Ocola, L. E., & Chen, J., (2011). Ultrafast room temperature NH_3 sensing with positively gated reduced graphene oxide field-effect transistors. *Chemical Communications, 47*(27), 7761–7763.

51. Lu, Y., Goldsmith, B. R., Kybert, N. J., & Johnson, A. C., (2010). DNA-decorated graphene chemical sensors. *Applied Physics Letters, 97*(8), 083107.

52. Mao, S., Chang, J., Pu, H., Lu, G., He, Q., Zhang, H., & Chen, J., (2017). Two-dimensional nanomaterial-based field-effect transistors for chemical and biological sensing. *Chemical Society Reviews, 46*(22), 6872–6904.

53. Mao, S., Lu, G., Yu, K., Bo, Z., & Chen, J., (2010). Specific protein detection using thermally reduced graphene oxide sheet decorated with gold nanoparticle□antibody conjugates. *Advanced Materials, 22*(32), 3521–3526.

54. Mcbride, P. T., Janata, J., Comte, P. A., Moss, S. D., & Johnson, C. C., (1978). Ion-selective field-effect transistors with polymeric membranes. *Analytica Chimica Acta, 101*(2), 239–245.

55. Meric, I., Han, M. Y., Young, A. F., Ozyilmaz, B., Kim, P., & Shepard, K. L., (2008). Current saturation in zero-bandgap, top-gated graphene field-effect transistors. *Nature Nanotechnology, 3*(11), 654–659.

56. Mohanty, N., & Berry, V., (2008). Graphene-based single-bacterium resolution biodevice and DNA transistor: Interfacing graphene derivatives with nanoscale and microscale biocomponents. *Nano Letters, 8*(12), 4469–4476.

57. Nolan, E. M., & Lippard, S. J., (2007). Turn-on and ratiometric mercury sensing in water with a red-emitting probe. *Journal of the American Chemical Society, 129*(18), 5910–5918.

58. Nomani, M. W., Shishir, R., Qazi, M., Diwan, D., Shields, V. B., Spencer, M. G., & Koley, G., (2010). Highly sensitive and selective detection of NO_2 using epitaxial graphene on 6H-SiC. *Sensors and Actuators B: Chemical, 150*(1), 301–307.
59. Novoselov, K. S., Geim, A. K., Morozov, S. V., Jiang, D., Katsnelson, M. I., Grigorieva, I., & Firsov, A. A., (2005). Two-dimensional gas of massless Dirac fermions in graphene. *Nature, 438*(7065), 197–200.
60. Novoselov, K. S., Geim, A. K., Morozov, S. V., Jiang, D., Zhang, Y., Dubonos, S. V., & Firsov, A. A., (2004). Electric field effect in atomically thin carbon films. *Science, 306*(5696), 666–669.
61. Ohno, Y., Maehashi, K., & Matsumoto, K., (2010). Label-free biosensors based on aptamer-modified graphene field-effect transistors. *Journal of the American Chemical Society, 132*(51), 18012–18013.
62. Ohno, Y., Maehashi, K., Yamashiro, Y., & Matsumoto, K., (2009). Electrolyte-gated graphene field-effect transistors for detecting pH and protein adsorption. *Nano Letters, 9*(9), 3318–3322.
63. Patolsky, F., Zheng, G., & Lieber, C. M., (2006). *Nanowire-Based Biosensors, 1*, 1–33.
64. Peng, N., Zhang, Q., Chow, C. L., Tan, O. K., & Marzari, N., (2009). Sensing mechanisms for carbon nanotube based NH_3 gas detection. *Nano Letters, 9*(4), 1626–1630.
65. Pumera, M., (2011). Graphene-based nanomaterials for energy storage. *Energy and Environmental Science, 4*(3), 668–674.
66. Radisavljevic, B., Radenovic, A., Brivio, J., Giacometti, V., & Kis, A., (2011). Single-layer MoS_2 transistors. *Nature Nanotechnology, 6*(3), 147–150.
67. Ramgir, N. S., Yang, Y., & Zacharias, M., (2010). Nanowire□based sensors. *Small, 6*(16), 1705–1722.
68. Ratinac, K. R., Yang, W., Gooding, J. J., Thordarson, P., & Braet, F., (2011). Graphene and related materials in electrochemical sensing. *Electroanalysis, 23*(4), 803–826.
69. Reddy, D., Register, L. F., Carpenter, G. D., & Banerjee, S. K., (2011). Graphene field-effect transistors. *Journal of Physics D: Applied Physics, 44*(31), 313001.
70. Ristein, J., Zhang, W., Speck, F., Ostler, M., Ley, L., & Seyller, T., (2010). Characteristics of solution gated field-effect transistors on the basis of epitaxial graphene on silicon carbide. *Journal of Physics D:*

Applied Physics, 43(34), 345303.

71. Robinson, J. T., Perkins, F. K., Snow, E. S., Wei, Z., & Sheehan, P. E., (2008). Reduced graphene oxide molecular sensors. *Nano Letters, 8*(10), 3137–3140.
72. Schedin, F., Geim, A. K., Morozov, S. V., Hill, E. W., Blake, P., Katsnelson, M. I., & Novoselov, K. S., (2007). Detection of individual gas molecules adsorbed on graphene. *Nature Materials, 6*(9), 652–655.
73. Schwierz, F., (2010). Graphene transistors. *Nature Nanotechnology, 5*(7), 487.
74. Stewart, M. E., Anderton, C. R., Thompson, L. B., Maria, J., Gray, S. K., Rogers, J. A., & Nuzzo, R. G., (2008). Nanostructured plasmonic sensors. *Chemical Reviews, 108*(2), 494–521.
75. Stine, R., Mulvaney, S. P., Robinson, J. T., Tamanaha, C. R., & Sheehan, P. E., (2013). Fabrication, optimization, and use of graphene field-effect sensors. *Analytical Chemistry, 85*(2), 509–521.
76. Sudibya, H. G., He, Q., Zhang, H., & Chen, P., (2011). Electrical detection of metal ions using field-effect transistors based on micropatterned reduced graphene oxide films. *ACS Nano, 5*(3), 1990–1994.
77. Sun, Y., Wu, Q., & Shi, G., (2011). Graphene based new energy materials. *Energy and Environmental Science, 4*(4), 1113–1132.
78. Sze, S. M., & Ng, K. K., (2006). *Physics of Semiconductor Devices* (Vol. 1, pp. 1–33). John Wiley & sons.
79. Wang, L., Tian, L. H., Wei, G. D., Gao, F. M., Zheng, J. J., & Yang, W. Y., (2011). Epitaxial growth of graphene and their applications in devices. *Journal of Inorganic Materials, 26*(10), 1009–1019.
80. Wang, Z., Zhang, J., Chen, P., Zhou, X., Yang, Y., Wu, S., & Zhang, Q., (2011). Label-free, electrochemical detection of methicillin-resistant *Staphylococcus aureus* DNA with reduced graphene oxide-modified electrodes. *Biosensors and Bioelectronics, 26*(9), 3881–3886.
81. Wang, Z., Zhang, J., Yin, Z., Wu, S., Mandler, D., & Zhang, H., (2012). Fabrication of nanoelectrode ensembles by electrodeposition of Au nanoparticles on single-layer graphene oxide sheets. *Nanoscale, 4*(8), 2728–2733.
82. Wehling, T. O., Novoselov, K. S., Morozov, S. V., Vdovin, E. E., Katsnelson, M. I., Geim, A. K., & Lichtenstein, A. I., (2008). Molecular doping of graphene. *Nano Letters, 8*(1), 173–177.

83. Willets, K. A., & Van, D. R. P., (2007). Localized surface plasmon resonance spectroscopy and sensing. *Annu. Rev. Phys. Chem., 58*, 267–297.

84. Wu, S., Yin, Z., He, Q., Huang, X., Zhou, X., & Zhang, H., (2010). Electrochemical deposition of semiconductor oxides on reduced graphene oxide-based flexible, transparent, and conductive electrodes. *The Journal of Physical Chemistry C, 114*(27), 11816–11821.

85. Wu, S., Yin, Z., He, Q., Lu, G., Yan, Q., & Zhang, H., (2011). Nucleation mechanism of electrochemical deposition of Cu on reduced graphene oxide electrodes. *The Journal of Physical Chemistry C, 115*(32), 15973–15979.

86. Wu, S., Zeng, Z., He, Q., Wang, Z., Wang, S. J., Du, Y., & Zhang, H., (2012). Electrochemically reduced single□layer MoS_2 nanosheets: Characterization, properties, and sensing applications. *Small, 8*(14), 2264–2270.

87. Wu, W., Liu, Z., Jauregui, L. A., Yu, Q., Pillai, R., Cao, H., & Pei, S. S., (2010). Wafer-scale synthesis of graphene by chemical vapor deposition and its application in hydrogen sensing. *Sensors and Actuators B: Chemical, 150*(1), 296–300.

88. Yamada, T., Letunic, I., Okuda, S., Kanehisa, M., & Bork, P., (2011). iPath2. 0: Interactive pathway explorer. *Nucleic Acids Research, 39*(2), W412–W415.

89. Yang, M., Zhang, S., Qu, F., Gong, S., Wang, C., Qiu, L., & Cheng, W., (2019). High performance acetone sensor based on ZnO nanorods modified by Au nanoparticles. *Journal of Alloys and Compounds, 797*, 246–252.

90. Yavari, F., Chen, Z., Thomas, A. V., Ren, W., Cheng, H. M., & Koratkar, N., (2011). High sensitivity gas detection using a macroscopic three-dimensional graphene foam network. *Scientific Reports, 1*(1), 1–5.

91. Yi, J., Lee, J. M., & Park, W. I., (2011). Vertically aligned ZnO nanorods and graphene hybrid architectures for high-sensitive flexible gas sensors. *Sensors and Actuators B: Chemical, 155*(1), 264–269.

92. Yin, P. T., Kim, T. H., Choi, J. W., & Lee, K. B., (2013). Prospects for graphene-nanoparticle-based hybrid sensors. *Physical Chemistry Chemical Physics, 15*(31), 12785–12799.

93. Yin, Z., Sun, S., Salim, T., Wu, S., Huang, X., He, Q., & Zhang, H. (2010a). Organic photovoltaic devices using highly flexible reduced

graphene oxide films as transparent electrodes. *ACS nano*, *4*(9), 5263-5268.

94. Yin, Z., He, Q., Huang, X., Zhang, J., Wu, S., Chen, P., & Zhang, H., (2012). Real-time DNA detection using Pt nanoparticle-decorated reduced graphene oxide field-effect transistors. *Nanoscale, 4*(1), 293–297.
95. Yin, Z., Wu, S., Zhou, X., Huang, X., Zhang, Q., Boey, F., & Zhang, H., (2010b). Electrochemical deposition of ZnO nanorods on transparent reduced graphene oxide electrodes for hybrid solar cells. *Small, 6*(2), 307–312.
96. Yu, K., Wang, P., Lu, G., Chen, K. H., Bo, Z., & Chen, J., (2011). Patterning vertically oriented graphene sheets for nanodevice applications. *The Journal of Physical Chemistry Letters, 2*(6), 537–542.
97. Zeng, Z., Yin, Z., Huang, X., Li, H., He, Q., Lu, G., & Zhang, H., (2011). Single□layer semiconducting nanosheets: High□yield preparation and device fabrication. *Angewandte Chemie, 123*(47), 11289–11293.
98. Zhang, J., Shen, G., Wang, W., Zhou, X., & Guo, S., (2010). Individual nanocomposite sheets of chemically reduced graphene oxide and poly(N-vinyl pyrrolidone): Preparation and humidity sensing characteristics. *Journal of Materials Chemistry, 20*(48), 10824–10828.
99. Zhang, T., Cheng, Z., Wang, Y., Li, Z., Wang, C., Li, Y., & Fang, Y., (2010). Self-assembled 1-octadecanethiol monolayers on graphene for mercury detection. *Nano Letters, 10*(11), 4738–4741.
100. Zhang, Y. H., Chen, Y. B., Zhou, K. G., Liu, C. H., Zeng, J., Zhang, H. L., & Peng, Y., (2009). Improving gas sensing properties of graphene by introducing dopants and defects: A first-principles study. *Nanotechnology, 20*(18), 185504.
101. Zhang, Y., Nayak, T. R., Hong, H., & Cai, W., (2012). Graphene: A versatile nanoplatform for biomedical applications. *Nanoscale, 4*(13), 3833–3842.

CHAPTER 5

Synthesis and Application of Graphene for Solar Cells

CONTENTS

5.1. INTRODUCTION

Graphene physical and chemical properties make it a suitable material for coming future energy-harvesting, optics, and also electronic devices. Graphene contains great optical transparency plus mobility together along with outstanding chemical inertness and mechanical characteristics. Single-layer graphene shows very high optical transmission (around 98%), also permits passing from the extensive scale of light wavelengths, therefore for an optically conducting window selected as an ever-studied material. Moreover, by applying different chemical functionalization procedures, graphene's electrocatalytic, electrical, and optical properties can be changed, which makes it one of the most useful applicants in energy-harvesting and optoelectronic devices for advanced applications. This chapter is planned to review the most significant exploratory outcomes from studies about characteristics of electrodes of graphene and its usage in several kinds of solar cells. Moreover, the development of various graphene functionalization and mixture processes for its utilization in the solar cells are as well explained in that particular chapter (Geim and Novoselov, 2010; Balandin, 2011).

We know graphene is a dense 2D material, a single atom, therefore, showing around 97.7% transfer through the whole spectrum of visible light. Moreover, graphene has a property of spectrum of smooth transmission from the UV, i.e., ultraviolet area to the elongated wave IR region, i.e., length infrared region, therefore showing an extensive space that permits the complete series of photon wavelength crossed via it. Absolutely, excluding graphene, these mixtures of excellent optical characteristics are having very useful in every kind of material until now. Likewise, graphene having rare electric carriage characteristics, through which we can track the features of Landau level quantization, quantum hall effects, 2D Dirac fermions, and go on (Novoselov et al., 2004). In further two dimensions, therefore, free charges in graphene are stable in just one longitudinal dimensional angle but movable in other, so therefore in free-standing graphene charge carrier movement is about 10^6 $cm^2/V/s$. Likewise, graphene also displays brilliant thermal and mechanical characteristics (k around 3000–5000 W/mK)[1] and it is also inert chemically. Therefore, characteristics among optical characteristics present together with graphene, which is an excellent applicant for uses in energy collecting devices, elastic optoelectronics, TCEs (transparent conducting electrodes), photodetectors, and several further optical devices. Moreover, graphene is predicted as a developing alternative for common TCEs, precisely, ITO (indium tin oxide), which comprises indium even as a minute component, expensive plus toxic. Specifically, in solar cells, the graphene-based (transparent conducting electrodes for the application with improved productivity is of greatest attention (Novoselov et al., 2004; Choi and Lee, 2011).

Up to now, for several kinds of solar cells, graphene electrodes have been useful, specifically, quantum dot, electrochemical, solid-state, and polymer solar cells. In different solar cells the key outcomes of using graphene are given here: (i) it generates the space for persuading extensive scales (starts from ultraviolet to the infrared regions) of the energy of photon within the solar cells, (ii) by the boundary of the electrochemical fusion cells that shows greater CT (charge transfer) kinetics, (iii) with strong architecture synthesis a flexible device, and (iv) it gives higher heat dissipation. Instead, increasing the productivity of electrochemical solar cells, the electrocatalytic performances of graphene play an important role such as DSSCs also used as "dye-sensitized solar cells," in which solid/liquid boundary is a path for transporting electronic charges. Nevertheless, due to the passive nature of the graphene basal at the liquid/graphene boundary tight CT is kept in spite of the great in smooth electron movement, therefore from the edge-planes improvement in electrocatalytic actions happens only. Therefore, for electrochemical solar cells, graphene has to be surface functionalized for enhancement of in-plane CT improve as well as uses (Viculis et al., 2003; Hernandez et al., 2008).

5.2. GRAPHENE SYNTHESIS: A BRIEF OVERVIEW

A single graphitic layer made of carbon nanostructures as its major structural blocks, and arises in a hexagonal honeycomb frame of covalently sp^2 bonded carbon atoms (Figure 5.1). The layers of a honeycomb-like graphitic matrix are bonded as well as stacked via a very weak van der Waals force makes three-dimensional (3D) bulk graphite. A sphere of a single graphite layer is called the zero-dimensional nanotubes, regarding its axis makes a 1D cylindrical assembly when rolled up, known as a carbon nanotube (CNT), and displays a planar two-dimensional (2D) structure which can have single or multiple layers bonded, and are termed as graphene. The single-layered graphene is the one having one graphitic layer and similarly, bilayer, and trilayer graphene are the ones having 2 and 3 graphitic layers, respectively. Above 5 layers to 10 layers, graphene is usually called the graphene of few-layer and similarly, multilayer graphene is the one having around 20 to 30 layers (Hummers and Offeman, 1958).

This chapter summarizes the major and commonly used processes for synthesizing graphenes, such as CVD, chemical methods, exfoliation, and the epitaxial growth composed for the uses in solar energy devices with a short discussion of their feasibility (Li et al., 2010).

5.2.1. Mechanical Exfoliation

In nanotechnology, mechanical exfoliation is a top-down manufacturing technique, through this, transverse or longitudinal stress on layered material's surface is produced by the use of either scotch tape or by using the tip of AFM to bare the required layers from materials. Although by using AFM tips, mechanical exfoliation approaches were in early 1999 tested, the method was ineffective to give a graphene layer with a thickness of one atom. This distinct kind of mechanical exfoliation method was implemented by Novoselov et al. (2004) through which they effectively produced a thick graphene layer of a single atom (Hasin et al., 2010).

5.2.2. Chemical Exfoliation

The alkali metal ions in this method, are employed to interpose the bulk graphite structure to isolate layers of graphene subsequently as a solution dispersion. The major motives in intercalation reactions including alkali metals are the following: (i) They contain a lesser atomic radius compared to the interlayer distance of graphite and therefore in the interlayer window they can easily fit through the reaction of intercalation, (ii) alkali metal ions can easily form intercalated structures by reacting with graphite. The main alkali metal ions are Na, Na-K_2 alloy, Cs, and Li which is usually for the chemical exfoliation process applied to interpolate graphite. Moreover, the exfoliation reaction's by-product, KC_8, as the reaction takes place with aqueous ethanol solution experiences an exothermic reaction according to Eqn. (1).

$$KC_8 + CH_3CH_2OH \rightarrow 8C + KOCH_2CH_3 + 1/2H_2 \quad (1)$$

Similarly, in organic solvents by applying the sonication method, the graphene's chemical exfoliation is also exhibited subsequently the dispersal. Hernandez et al. (2008) by use of a straightforward sonication method described the exfoliation of pure graphite in N-methyl pyrrolidone (NMP), which produces good quality unoxidized monolayer graphene (Viculis et al., 2003).

5.2.3. Chemical Synthesis and Functionalization

There are few steps involved in the chemical synthesis of graphene-like then reducing it back to graphene, in a solvent the dispersal of layers of graphene-oxide, and oxidation of graphite. By using Hummer's method, the oxidation of graphite is occurred by the reaction of graphite with potassium permanganate, sodium nitrite, and sulfuric acid (Hummers and Offeman, 1958). The aim of converting graphite into GO (graphite oxide) is the growth of the interlayer window greater than the original graphite. Proportionally the interlayer spacing differs in this regard to the extent of graphite's oxidation. This graphite oxidized further enables the dispersal of graphene in suitable solvents. In a polar liquid medium, GO (graphene oxide) is generally dispersed, for instance, NMP, dimethyl sulfoxide (DMSO), hexamethyl phosphoramide (HMPA), dimethylformamide (DMF). Moving on, at 80°C, the dispersed GO is transformed into graphene by applying dimethylhydrazine (DMH) treatment for a long time. The homogeneousness of GO distribution decisively differs in regards to kinds of functional groups grafted to graphene, that in the solution more grip the graphene accumulation. Therefore, homogeneous dispersal gives rise to the homogeneous reduction of GO with improved properties. Now, used for organic solar cells, numerous reports have been exhibited on the uses of QDSCs, DSSCs, GO, Li-ion batteries, organic memory devices, and several more. The main uses of the chemical synthesis method are solution-based and low-temperature procedure, and so it owns great flexibilities on numerous substrates used for direct production plus scalability of graphene film. Furthermore, graphene's in situ workings can be simply achieved by applying this procedure for catalytic possessions and change of graphene's chemical. Though, the chemical production method of graphene production has numerous limitations, for example, (i) defective structure, (ii) minor yield, and (iii) partial decrease of graphene, which gladly damages the graphene characteristics (Hummers and Offeman, 1958).

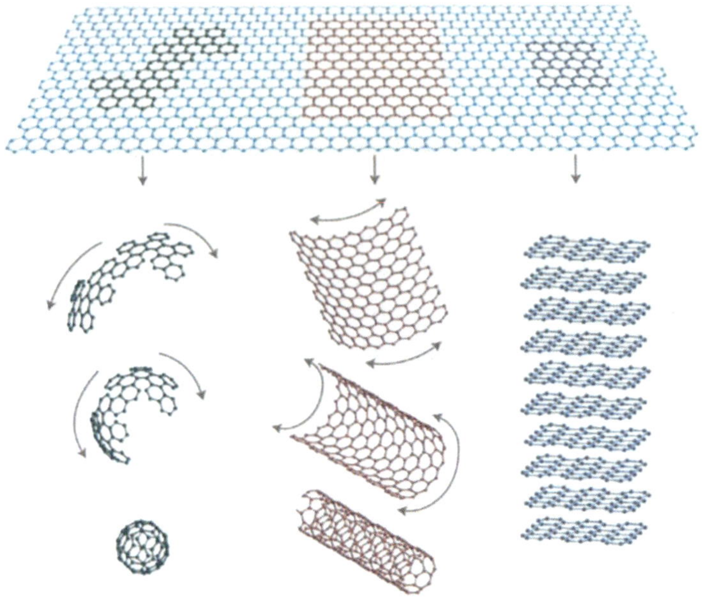

Figure 5.1. Shows two-dimensional graphene hexagonal sheets as the structure of further types of CNs.

Source: https://www.cambridge.org/core/journals/journal-of-materials-research/article/graphene-synthesis-and-application-for-solar-cells/A67526BEB-97F96054E65F723A9F751D2/core-reader.

5.2.4. Thermal CVD Process

At high-temperature graphene is formed on Cu and Ni by applying the chemical vapor deposition (CVD) of hydrocarbon gases in condensed atmospheric conditions. In this process, on the surface of transition metals as substrates Ni and Cu foils are applied, where graphene deposition happens because of the catalysis method. Numerous metals like Ir, Co, Fe, Pt, Pd, and Ru are for the graphene's deposition also used as substrates via CVD (Li et al., 2010). Around the 1000°C, the process temperature was reserved (even though numerous lower or greater temperature methods were also described) below decreasing H_2 atmosphere to decay methane and yield graphene according to the given Eqn. (2):

$$H_2 + CH_4 = 2H_2 + C \quad (2)$$

The procedure is scalable and useful as the catalyzing method does not differ with the substrate dimensions. Though by the CVD method, large-scale graphene can be attained, however, attaining a high scale graphene's homogeneous layer is yet under trial. Furthermore, during CVD methods in graphene, the creation of ripples and grain boundaries can result in defects in graphene. These defects produce key charge-scattering, also decline the thermal, optical, and electrical characteristics of graphene (Hasin et al., 2010).

5.2.5. Epitaxial Growth

For producing very crystalline graphene, the epitaxial growth method is performed on single crystal-like silicon carbide (SiC) substrates. Mono-crystal graphene can be produced by sublimation of silicon through 6H-SiC. The growth method consists of surface preparation by using H_2 etching or oxidation, after that by the virtue of bombarding electrons surface cleaning at 10^{10} Torr pressure at 1000°C for 1–20 min below at 1250–1450°C the heat treatment (Li et al., 2010). At high temperature on the Si sublimation process, this growth method is based, which in graphene makes the anisotropy. Depending on the thermal decomposition process and several other factors on SiC the epitaxially grownup graphene shows larger and lesser domain size-based. Through this method the great quality higher score epitaxial graphene is also attained, but, the transmission of graphene via SiC to further substrates is tough, and the method is also costly. So, for solar cell applications, these extremely controlled the epitaxially full-grown graphene (Roy-Mayhew et al., 2010).

5.2.6. Graphene Transfer

The process in which fabricates the graphene-based photovoltaic devices occur is known as the graphene transfer as graphene is manufactured a thin film on the metal foils or in a discrete solution form. Several methods have now been established to transfer graphene on further substrates, for instance, flexible substrates or glass. Some few common approaches of transfer of graphene from the discrete medium to a substratum are the casting of the drop, electrophoretic deposition, dip casting, and spin coating. Furthermore, some other methods, like Langmuir-Blodgett (LB) assembly, spraying, vacuum filtration, and self-assembly approaches, are useful for transferring graphene on several flexible and transparent substrates. In contrast, the transfer of graphene procedures from foils of metals towards further substrate materials are roll-to-roll transferring process, stamping process, lamination hot press method, PDMS transfer method, and chemical transfer procedure (Chen and Wang, 2010).

5.2.7. Graphene as Transparent Conducting Electrodes

Generally, transparent conducting electrodes, keeping resistivity in control is one of the problematic challenges that is importantly based on the 2 material factors of great carrier mobility and great carrier concentrations. Practically, the common transparent conducting oxide (TCO) materials were having some drawbacks, like large bandgap, complex structure, and inevitable charge scattering centers (which is rising from the existence of basic dopants, defects, etc.), and were restraining their action. As compared to general TCEs in the background of research exceedingly transparent conducting electrodes with extraordinary carrier movement, single layer graphene-based transparent conducting electrodes are becoming very common (Wintterlin and Bocquet, 2009). The carrier mobility as well as a zero-band gap (numerous orders greater than the common TCOs) gives great applicability like TCEs in nanoelectronics. Moreover, numerous scalable and simple methods for useful transfer of graphene on a polymer substrate have been established, which is carefully comprehensive and technically important for transparent, elastic devices. Geim and Novoselov (2010) described that transparency mono-layered graphene is about 98%. Nevertheless, sheet resistance plus transparency reduces with the growing amount of graphene layers as the following: trilayer graphene displays the 700 X/sq sheet resistance and 92% transmission, bilayer graphene shows 1 KX/sq sheet resistance and 95% transmission, and 4th-layer graphene owns 400 X/sq sheet resistance and 89% transmission. Wang et al. (2013) exhibited the translucent electrodes of graphene having transparency of 80%, as a TCO of DSSCs, which is shown. Verma et al. (2010) described the high-scale transference on the polymer of few-layer graphene with the use of a common lamination of hot press technique having transparency of 89%. They represented a less-cost hot press and easy technique that was stretched as big as 15 x 5 cm. Polymer film with graphene was used as an anode for great resolution elastic field emission (FE) devices, with a stretch where elastic graphene exhibited fixable electrical characteristics. Likewise, Lahiri et al. (2011) also exhibited the flexible and complete transparent FE device containing both graphene-based anode and cathode. In that background, for the production of large graphene, a roll-to-roll method has been shown for touch screen uses. Bae et al. (2010) on a polymer described the transmission of mono-to-four-layer graphene film through the transparency of about 97%

to 90%. In a similar description, it is made clear that the resistance of the sheet of the film of graphene as well diverse from 275 to 50 X/sq for 1 to 4th-layered graphene (Berger et al., 2004; Hupalo et al., 2009).

5.3. GRAPHENE IN SOLAR CELLS

5.3.1. Graphene for Solid-State Solar Cells

Even though graphene is a close to null band-gap material, because of the electrical conductivity and high mobility shown by graphene, the connection of graphene/n-type semiconductor heterojunction can be taken to be effective Schottky intersection. In order to make a built-in potential, it is important to mention here that the work-function variance among the n-type semiconductor and graphene should be sufficient as presented in Figure 5.2(a). Figure 5.2(a) shows the band configuration demonstrating along with a built-in probable of ($u_n u_{G-Si}$) the theme of an n-type Schottky/ graphene intersection solar cell. In this background, only graphene-single CdS NW (nanowire) Schottky intersection solar cells and graphene-Si have been described up till now (De Heer et al., 2007; Emtsev et al., 2009).

Li et al. (2009a, b) was the first to report the n-Si solar/graphene cell, also, device assembly is visible in Figure 5.2(b). Initially, via etching the 300-nm SiO_2 layers among a wafer, they formed a channel by photolithographic modeling followed. By applying sputter deposition of Ag/Pd/Au and Ti metal contacts were formed on the front side and the backside of n-Si, (Figure 5.2(b)). Few layers of CVD graphene were moved through the set pattern to make the graphene/n-type Si Schottky intersection. Through the solar cells, a J_{SC} (current density of short-circuit) of 4.0–6.5 mA/cm^2, and a V_{OC} (open circuit voltage) of around 0.42–0.48 V along with a PCE (power conversion efficiency) of about 1.5% below the standard form: global (1.5G) glitter of 100 mW/cm^2, AM (air mass) 1.5 is exhibited. Likewise, some other publications were found citing graphene/Si nanopillar collection and graphene/Si nanocrystal Schottky intersection solar cells, where the stated PCEs are around 1.96% and 0.02 (Ye et al., 2010). In a similar publication, Berger et al. (2004) exhibited a technique of good development of the Si nanopillar array/graphene Schottky junction solar cells through the p-type doping of graphene by applying HNO_3 treatment. By applying various dopant concentrations, the p-type doped graphene Schottky intersection solar cells displayed proficiency around 3.55%. Miao et al. (2012) presented the improvement of graphene/Si Schottky intersection solar cell productivity of around 8.6%, which is the maximum stated solid-state solar cell of graphene-based productivity. Graphene is doped here accompanied by bis(TFSA)amide (trifluoromethane sulfonyl), which function as a hole dopant in graphene improving the work-function deprived of altering its optical properties (Becerril et al., 2008). The doping perused a change in the chemical potential of graphene responsible for causing the rise in the device's inherent potential along with carrier density upsurge in graphene and, therefore, increases the solar cell 6 around 8.6% (Miao et al., 2012). Particularly, Ye et al. (2010) by the use of a junction of CdS nanobelt (NB) or CdS NW by graphene described a novel form of Schottky intersection solar cell. Figure 5.3(a) displays in the graphene-CdS the diagram of Schottky intersection creation. The CdS NB and CdS NW were produced by the CVD technique and moved them onto a Si/SiO_2 substrate as represented in the equipment's infrastructure in Figure 5.3(b). The device displays a PCE around 1.65% (beneath AM 1.5G glitter) with equivalent FF, V_{OC}, and J_{SC} of 40%, 0.15 V, and 275 pA, respectively (Kymakis et al., 2011).

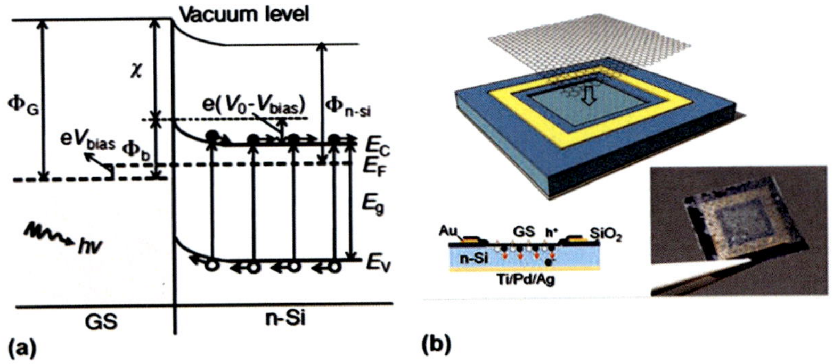

Figure 5.2. (a) Shows at the graphene/n-Si Schottky junction the band diagram; (b) diagram design of production method of graphene/n-Si Schottky intersection solar arrays.

Source: https://pubmed.ncbi.nlm.nih.gov/20379996/.

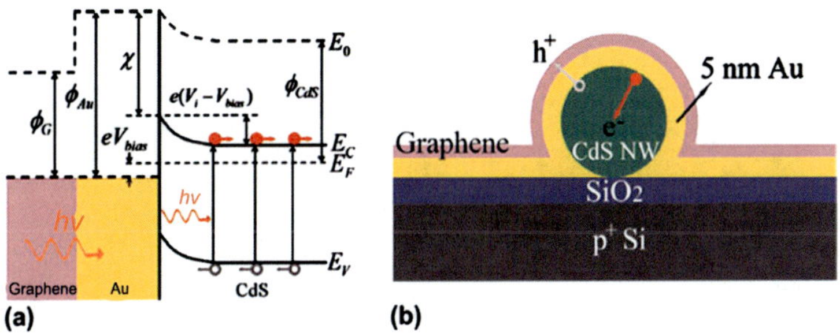

Figure 5.3. (a) The diagram representation of a graphene/CdS NW Schottky connection solar cell.(b) The band drawing of the Schottky junction of graphene/CdS NW

Source: https://pubs.acs.org/doi/10.1021/am1007672.

5.3.2. Graphene-Based DSSCs

5.3.2.1. Dye-Sensitized Solar Cells

The 3rd gen photo-voltaic cells such as DSSCs, OPVs (organic photovoltaics), and SQD (semiconductor quantum dots) have great respect in the circle of photovoltaic research (Mora-Seró and Bisquert, 2010; Tongay et al., 2011). In the field of produce short energy repayment time and roll-to-roll production, the new types of photovoltaic devices are applicable. In this chapter, we reviewed the developments on new graphene-based materials for the 3rd generation PVs research such as organic photovoltaics, QDSCs, and DSSCs (Wang et al., 2008; Chavez-Valdez et al., 2013).

Just, DSSCs have great attention as the capable incarnations in photovoltaic technologies for making affordable, clean, plus renewable energy. The use of nontoxic, low-cost components finds the DSSC for combined "green" designs as a perfect photovoltaic idea (Eda et al., 2008). In DSSCs, charge carriage and light absorption are detached with charge carrier group happening by the chemisorption of sensitizer molecules which forms self-assembled single layer inserted among photoelectrode which is a Semi-conducting anatase Titania) with electrolyte making the hole as well as electron conducting device (Figure 5.4(a)).

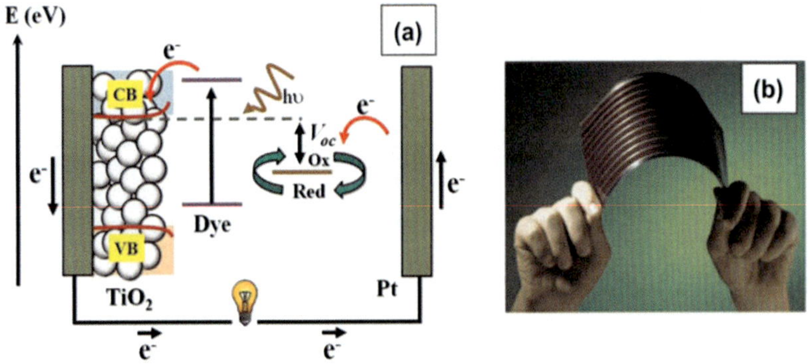

Figure 5.4. (a) DSSCs schematic diagram; (b) sandwich elastic DSC part.

Source: https://pubs.acs.org/doi/10.1021/cr900356p.

Beneath irradiation, the molecular dye photoexcitation guides the injection of the fast electron to the conduction band (CB) of Titania or other SC material. subsequently, photoelectrode electron transfer, whereby a novel state of dye is restored by the acceptance of electrons from the electrolyte. In the photoelectrode, the collected electrons are carried to the Counter Electrode, via an outward circuit and renewing electrolyte the circuit is finished at CE (Li et al., 2008). Now, the study has enhanced on improving counterparts (sensitizers, electrolyte, photoanodes, also CEs) to improve the stability and efficiency of DSSCs to carry into the market (Figure 5.4(b)).

5.3.2.2. Graphene as Photoanode

In recent years, due to their unique features graphene encouraged a great deal of attention in DSSCs like (i) changeable electrochemical properties, (ii) excellent light transmittance ability, also (iii) fast electron transport. On the graphene-based DSSCs, several types of research have been showing, and in Figure 5.5 they are shown. Wang et al. (2013) discovered graphene as the good transmission electrode substance (up to 97.7%), in DSSCs for using graphene it is the initial effort that unlocked a new path (Figure 5.5(a)). In short, photoanode works as a charge automobile to carry the charges from dye sensitizers when excited to the outside circuit. In this process, TiO_2-Graphene is a very good material from which electron transfer is present between Titania and current collector (Kim et al., 2007). The band spot of graphene is the purpose behind it [NHE (normal hydrogen electrode) has 4.4 eV], which occurs among the FTO (fluorine-doped tin oxide) TiO_2 (NHE 4.2) and (NHE 4.7 eV). Therefore, it is expected the insertion of graphene among TiO_2 and we will get a working CT path to photo-anode by FTO. In comparison with the rest of 1-D composite implementors, spatially good discrete graphene sheets outspread great viability with nanocrystalline Titania particles. Furthermore, graphene works as a secondary binder (up to 0.8 wt.%) in the creation of free of crack Titania mesoporous dense films deprived of cooperating the photoconversion productivity of the device (Hagfeldt et al., 2010; Yella et al., 2011).

During the synthesis of TiO_2-Graphene (G-T composite) different types of methods have been exhibited. The GT composite can be fabricated through: (i) on TiO_2 framework decoration of graphene (Figure 5.6(a)) also (ii) Nanoparticles deposition by TiO_2 in-situ deposition on the graphene scaffold (Figure 5.6(b)). By using both TiO_2 outline or pre-produced graphene matrix, the G-T composite is attainable via either of the mentioned methods: electrospinning, spin coating, microwave-assisted, electrophoretic depositions, as well as hydrothermal (Chen et al., 2013). Generally, the G-T composite improves the J_{SC} (photocurrent density) of DSSCs in comparison with simple Titania electrode (Figure 5.6(c)). Improvement in photocurrent density is ascribed to (i) the visual scattering effect, from solar spectrum which guides to red-photon gathering and (ii) at TiO_2/dye interfaces enabled charge transport from graphene thruway, which at the photoanode helps the charged gathering (Tu et al., 2013; Wang et al., 2013). In research work, few papers described that the G-T composites improve the photocurrent density at TiO_2/dye interfaces via dropping the charge collection; though, Song et al. (2011) proved that the CB of Titania cannot be shifted by graphene and neither can it affect the recombination procedure. So, photocurrent density improvement is generally related to scattering and light absorption properties in the region that are visible, of graphene flakes and secondary transport along with the TiO_2 (graphene shows unusually great electron movement below normal conditions 15,000 $cm^2/V/s$). However, in G-T photoanode, composite quantity (wt%) is one of the greatest important features, which affects directly the device PCE. This is generally known that the addition of graphene more than the improved wt.% in TiO_2 results in a drop in the device performance. Although, in the wavelength scale of around 200–800 nm graphene is identified as to be transparent, but, into the TiO_2 matrix, the more quantity of graphene drops the light obtainability for dye molecules, which from dye sensitizers eventually reduces the light-harvesting influence (Cote et al., 2009; Ishikawa et al., 2012).

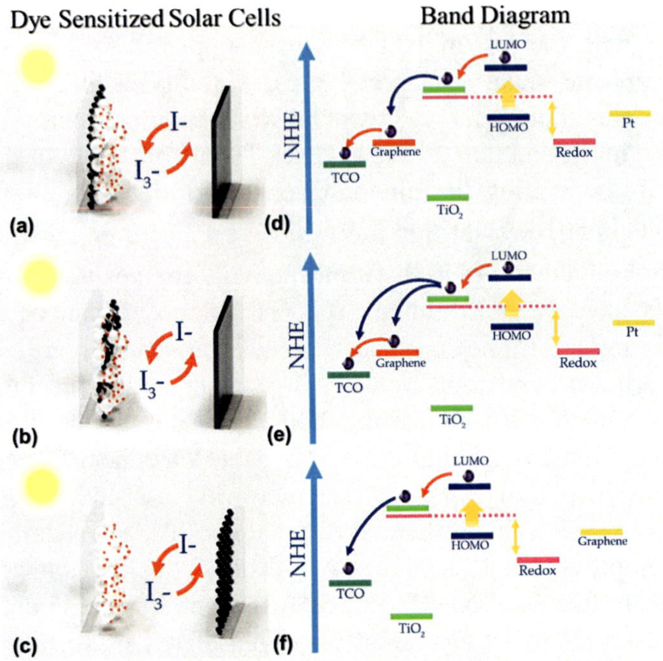

Figure 5.5. Three dissimilar DSSCs in (a) graphene shown as TCO; (b) as the anode; (c) also a cathode; (d), (e), and (f) represent the respective band figures of DSSCs shown in figure a, b, and c.

Source: https://www.cambridge.org/core/journals/journal-of-materials-research/article/graphene-synthesis-and-application-for-solar-cells/A67526BEB-97F96054E65F723A9F751D2.

In contrast, due to the graphene overfilling results in dye-loading quantity reduction in the G-T electrode occurs typically, thin G-T electrode pore size strictly restricts this loading amount. To counter this problem, common 2-Dimensional G-T composite conductors are changeable with 3-Dimensional extensive pore structure produced by applying 3D nickel foam or polystyrene inverse opals-scaffolds (Verma et al., 2010; Wang et al., 2013). Both electrolyte diffusion and high dye loading are being offered by these 3D GT electrodes. In these 3-Dimensional GT designs, graphene is combined nearby to the upper layers of the inverse opal structures afterward, fixed onto the TiO_2 matrix through post-treatment of the TiO_2 pioneers (Figure 5.6(d)).

It is detected that in 3D GT, the common procedure of graphene sheets efficiently improves the electron transport deprived of decreasing light-harvesting contribution via dye molecules. By the comparison of 2-D and 3-D G-T electrodes, this appeared that the 3-D G-T composite increases V_{OC} and J_{SC} contrary to 2-D G-T electrodes (Kim et al., 2013; Zhang et al., 2013). At extensive-pore natured photoanodes, the extraordinary V_{OC} improvement is attributed to the great electrolyte diffusion, which at TiO_2/dye interfaces eventually decreases the collection rate and increases the energy transfer places and therefore in these devices supports increased V_{OC} (Chen et al., 2009; Zhu et al., 2009).

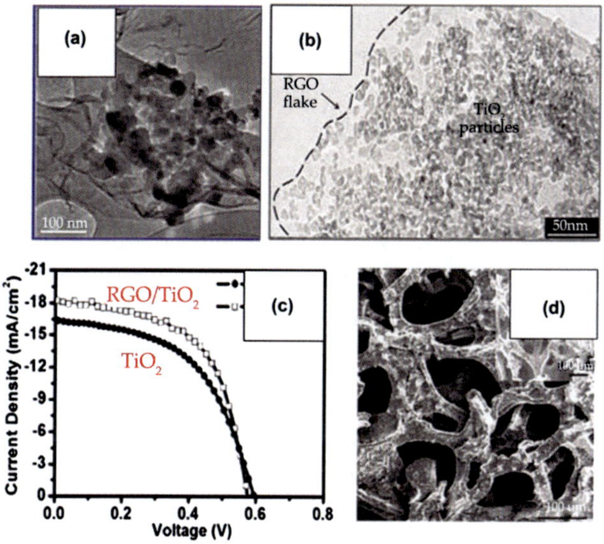

Figure 5.6. From dissimilar routes, G-T composite TEM images are shown: (a) graphene TiO_2; (b) TiO_2 graphene; (c) With rGO/TiO_2 and TiO_2-based photoanode, a comparative photovoltaic action of DSSC prepared; (d) by using Ni foam SEM images of 3-D G-T composite.

5.3.2.3. As Cathode Materials

C-E is an important component that is used to catalyze triiodide reduction (I_3 to I) inserts electrons into the electrolyte in DSSCs, after charge injection from the photo-oxidized dye. At the CE, the electrocatalytic reduction from I_3 to I [as shown in Eqn. (3)] injunctions the cathodic action of DSSCs and impacts charge production at the photo-anode via dye regeneration (Gong et al., 2013).

$$I_3^- + 2e^- \rightarrow 3I^- \quad . \tag{3}$$

For DSSCs, Pt (Platinum), due to its low resistance and excellent catalytic activity, is usually used as CEs. Nevertheless, for DSSCs, because of low chemical inertness, low abundance, and high cost, replacing platinum with other materials is being an essential problem. Up to now, in DSSCs, this is stated that the various types of materials containing carbon are relatively constant as well as best replacements for Pt CE (Das et al., 2011; Li et al., 2012). For having the potential to become CE material for solar cells, graphene has numerous advantages: ultrahigh transmittance, high thermal conductivity, high electrical conductance, and outstanding mechanical properties. In the whole solar spectrum, especially in the IR region, graphene has shown high transparency. So, graphene CEs are very beneficial for the kinds of tandem solar cells, which are required to be able to absorb a variety of photons (beginning from Visible-UV to IR), which are necessary to produce excitons proficiently (Kim et al., 2010; Li et al., 2012). Moreover, for attaining high surface area in a 2-D graphene sheet (2630 m^2/g), as well as its charge mobility and high value of intrinsic transmittance, DSSCs offer greater flexibility for a CE (Li et al., 2009b; Kim et al., 2010).

Hasin et al. (2010) were the first researchers who found the tri-iodide reduction ability of graphene, therefore, the applications from DSSC CEs shown extensive possibilities. According to this report, the functionalization of the surface through cationic polymer is a good method in improving graphene film's electrocatalytic property, having several advantages over increasing DSSC competence. The graphene nanoplatelet-based DSSC CE is shown by Kavan et al. (2011a) as an optically lucid cathode, which for the $I_{3/}I$ reaction is electro catalytically active. By applying a drop-cast technique they made the graphene-based cathode, and a PCE of around 5% is exhibited by the device. Roy-Mayhew et al. (2010) showed the functionalized-graphene sheet as DSSC CE along with a PCE of 3.83%. They showed the improvement in the electrocatalytic performance in graphene via functionalization of it by using poly(ethylene oxide) triblock poly(propylene oxide)-poly(ethylene oxide) copolymer. In this regard, Das et al. (2011) demonstrated employing fluorine ions among increased catalytic activity to the tri-iodide reduction in comparison with natural graphene the functionalization of large-scale CVD graphene. The PCE of the DSSC displayed; 2.56% with equivalent V_{OC}, FF, and J_{SC} of 0.66 mV, 35.9%, and 10.9 mA/cm^2, (Figure 5.7(j)–(n)). Likewise, HNO_3-doped graphene exhibited improvement in electrocatalytic performances in graphene along with a DSSC productivity of 3.21% as exhibited in Figure 5.7(f)–(i). Because of lower FF, the performance of the cell is lesser than predicted, which on the graphene/FTO interface creates a potential voltage drop. Hence, this is exhibited that functionalized graphene is very much operative compared to original graphene to improve at cathode the kinetics of I_3reduction. Graphene's electrocatalytic performance greatly differs from its functional groups containing oxygen and defect concentration as can be seen in Figure 5.8. Likewise, electrocatalytic actions by graphene are well related to its active spots, which at the interface play an important part in CT. Some reports presented that linkage of further functional groups like -NHCO-, -CO-, COOH, in graphene successfully increases graphene's electrocatalytic along with electronic properties. Just now, Lee et al. (2012a) for DSSC CEs suggested a novel graphene-based 3-D architecture as showed in Figure 5.9. 3-D doped GO foam in his report along with N and applied as a CE of DSSC, that exhibited a PCE of 7.07% (Bae et al., 2010; Song et al., 2011).

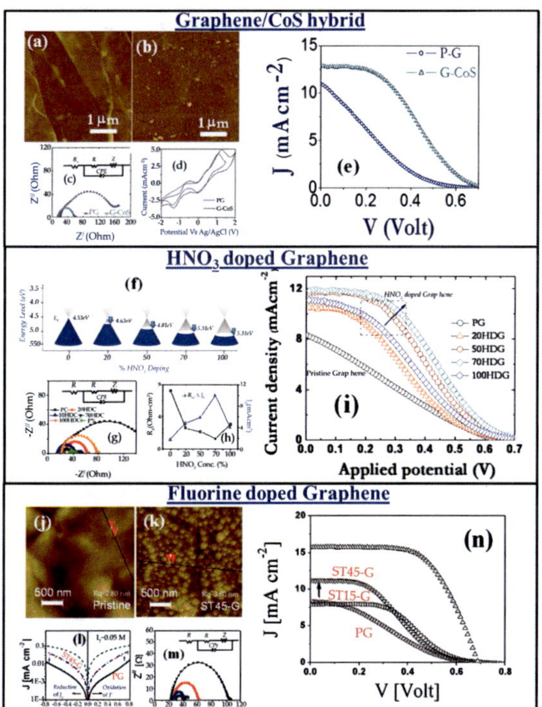

Figure 5.7. For DSSCs graphene synergetic CEs. (a–e) graphene-Co's mixture electrodes; (f–i) HNO_3-doped graphene; (j–n) Fluorine-doped graphene.

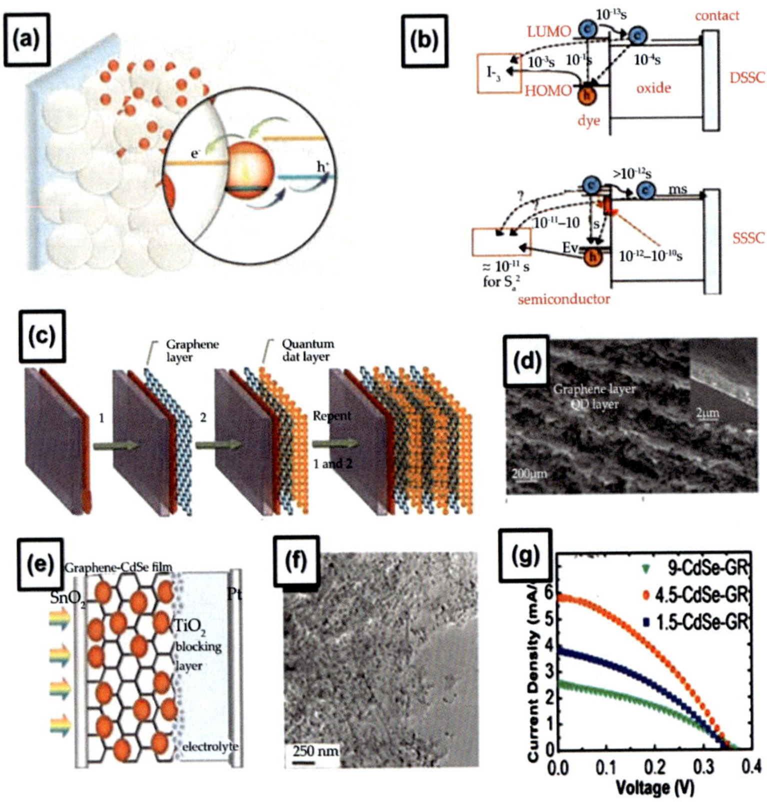

Figure 5.8. (a) at QD-sensitized TiO_2 interfaces representation of CT; (b) Associating CT times for the DSSC (top) and SSSC (bottom); (c) on ITO glass creation of the layered graphene/QDs; (d) cross-sectional SEM image of QD gathering is among the conformal graphene layers; (e) in QDSCs representation structure of QD-assembled graphene employing anode layer; (f) TEM images of the CdSe QD-decorated graphene matrix; (g) CdSe QD/graphene JV act in QDSCs.

Source: https://aip.scitation.org/doi/10.1063/1.3558732.

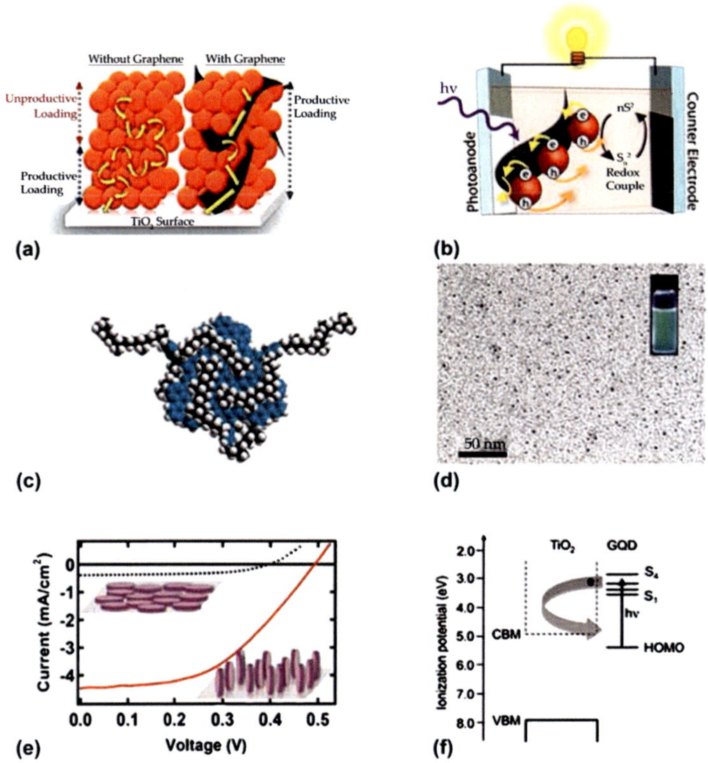

Figure 5.9. A schematic drawing of (a) in QD films comparing electron transport through and deprived of graphene layer; and (b) quantum dots-graphene composite created QDSCs; (c) enclosure 3D of the GQD core through the alkyl chains; (d) TEM images of green-luminescent GQD equipped by electrochemical oxidation of a graphene electrode in phosphate barrier solution, and the insert is a photo of a GQD aqueous solution beneath UV irradiation (365 nm); (e) J-V performance of GQD-sensitized TiO_2 QDSCs through two different GQD locations (GQDs signified as circular discs); (f) energy level diagram of the C132A molecule (GQD) on the TiO_2 surfaces (VBM, valence band maximum; CBM, CB minimum).

Source: https://pubs.acs.org/doi/abs/10.1021/nn305080c.

Similarly, for electrocatalytic cathode used for DSSC graphene fixed with polymer also exhibited promises. Numerous conducting polymers have been applied to improve the conductivity of the polymer/graphene composite along with the electrocatalytic activities such as polyaniline (PANI): poly (styrene sulfonate) (PEDOT: PSS), poly(3,4-ethylene dioxythiophene).

Hong et al. (2013) CE for DSSCs by 4.5% PCE described the PEDOT: PSS/graphene composite. In the report, researchers exhibited the increase of J_{SC} of DSSC after the addition of a slight quantity of graphene in PEDOT-PSS, which at the CE was credited to the greater catalyzing capability of graphene for the iodine reduction. Likewise, the PEDOT/graphene composite elastic electrode was shown by Lee et al. (2012), who made the graphene/PEDOT CE using an elastic PET substrate then gathered as DSSC (Chen et al., 2008; Kim et al., 2012).

Without Pt and TCO the PCE of DSSC is about 6.26%, which is nearly similar to DSSC with Pt and TCO (around 6.68%). Likewise, MnO_2-PANI/r-GO composite exhibited a PCE of around 6.15% along with FF, J_{SC}, and V_{OC} of about V65%, 12.88 mA/cm^2, and 0.74 m. One method that displayed the poly(diallyldimethylammonium) graphene/chloride oxide (graphene/PDDA) system was applied as DSSC CE along with around 9.5% PCE. Following this description, Gong et al. (2013) made the PDDA (cationic polymer) ornamented on the negatively charged GO by the process of layer-to-layer assembly. It resulted in improved catalytic performance in GO because of the charge difference among the nitrogen of ammonium ions and the electronegative carbon atoms (Akturk and Goldsman, 2008; Zhang et al., 2012).

In contrast, NP/graphene electrodes specify the research built on the composite cathode to great efficiency DSSC. In it, NPs attached to graphene as a CE for DSSCs have also been shown. In DSSCs for increasing graphene's electrocatalytic performance for triiodide reduction NP Pt-anchored graphene was exhibited. The main aim of these studies was to decrease the platinum load in DSSC CE by joining Platinum/graphene composite electrodes deprived of effecting its PCE. One study was found in graphene/silver NW composite electrodes including a PCE of almost 1.61%. Nonmetallic NPs like $Ni_{12}P_5$, NiO TiN, CoS has been applied to make graphene centric composite electrodes. CVD graphene/CoS electrodes presented a PCE of around 3.42% with FF, J_{SC}, and V_{OC} of; 36.40%,13.0 mA/cm^2, and, 0.72 V respectively (Morales-Torres et al., 2012; Chavez-Valdez et al., 2013).

Das et al. (2011) showed on graphene the grafting of CoS NPs improved incentive performances of electrodes of the composite because of the creation of catalytically active triple connection spots, at DSSC cathode which more helps the CT reactions (Figure 5.7(a) and (e)). Likewise, Chen et al. (2009) displayed the TiN/graphene NP composite electrodes for active dye-sensitized solar cell cathode having a PCE around 5.78%. In contrast, graphene/CNTs also exhibited high potentials for triiodide decrease in the dye-sensitized solar cell as the catalytic electrodes, thus, exhibited in new studies. Table 5.1 shows the specifics of the activities of DSSCs grounded on graphene-derived and CE graphene (Lee et al., 2012b; Zhu et al., 2012).

Table 5.1. Features of DSSCs based on Graphene CEs

	Synthesis Process	**FF (%)**	**PCE (%)**	**V_{OC} (V)**	**J_{SC} (mA/cm^2)**
Graphene CEs Graphene nanoplatelets	Commercial process	0.52	5.00	0.724	13.10
Graphene	Chemical synthesis procedure	36.0	0.26	0.7	1.01
GO (Graphene oxide)	Staudenmaier method	16.0	0.74	0.70	6.42
Graphene	Thermally exfoliated	54.0	2.82	0.68	7.70
rGO	Chemical synthesis process	54.0	6.81	0.75	16.99
rGO	Modified Hummer's method	56.0	2.19	0.64	6.12
rGO	Chemical synthesis process	65.0	5.69	0.54	14.30
rGO	Modified Hummer's method	46.0	2.64	0.72	8.11
Functionalized graphene CEs oxygen-functionalized graphene	Chemical synthesis	70.0	3.83	0.71	7.77
Fluorine-doped graphene	CVD process	35.9	2.56	0.66	10.9
NO_3-doped graphene	CVD process	37.0	3.21	0.73	11.8
-NHCO-functionalized r-GO	Chemical process	31.0	2.45	0.65	5.75
Nitrogen-doped graphene foam	Chemical process	58.0	7.07	0.77	15.84
Graphene composite CEs Graphene-CoS	CVD process	36.40	3.42	0.72	12.80
Graphene-Pt	Chemical synthesis	52.31	3.89	0.662	11.25
Graphene/Pt	Thermal exfoliation/chemical synthesis	59.0	2.91	0.74	6.67
Graphene-Pt	Chemical synthesis	71.0	7.66	0.71	15.20

Carbon-graphene nano-composite	Chemical synthesis	77.0	9.69	0.846	7.60
Graphene-$Ni_{12}P_5$	Hydrothermal reaction process	61.0	5.70	0.727	12.86
rGO-NiO	Chemical synthesis	61.0	3.06	0.67	7.53
rGO-CNTs	Chemical synthesis	62.7	7.55	0.75	16.05
rGO-CNTs	Chemical synthesis	70.0	3.00	0.76	5.60
Graphene-MWCNT	CVD synthesis	70.0	4.46	0.72	8.95
r-GO-MWCNT	Chemical process/CVD grown	62.7	7.55	0.75	16.05
r-GO-MWCNT	Chemical synthesis	61.3	6.17	0.78	12.86
Graphene-PEDOT	Chemical synthesis	63.0	6.26	0.77	12.60
Graphene-PEDOT-PSS	Chemical synthesis	48.0	4.50	0.72	12.96
Graphene oxide and positively charged poly(diallyldimethy lam-monium chloride)	Chemical synthesis	74.0	9.54	0.692	18.77
rGO-PANI	Chemical synthesis	65.0	6.15	0.74	12.88
Graphene-Ag NW	Commercially available	0.52	1.61	0.55	6.45
Graphene-TiN	Chemical synthesis	64.33	5.78	0.728	12.34

5.3.3. Graphene-Based Quantum Dot-Sensitized Solar Cells

Besides dye-sensitized solar cells, SQD as the light captivating materials in QDSSC (QD-sensitized solar cells) also bring attention through their attractive characteristics. Considering to change the absorption spectrum of SQD, examine the particle magnitude is an effective method to produce the complete series of the solar spectrum. Furthermore, due to the particular structure of the electronic band, SQD can control the Queisser-Shockley threshold of efficiency of energy conversion (Tsai et al., 2011; Madhavan et al., 2012). The capability of SQD to produce high-temperature electrons and to create manifold carriers forms them intended for the light-harvesting sensitizers in solar cells a feasible candidate. Numerous semiconductor materials (PbS, CdSe, CdS, etc.), have been functioned employing light sensitizers on mesoporous metal oxide layers of extensive bandgap (ZnO and TiO_2) because of its simple sensitization processing and low cost. The working of quantum dot solar cells is similar to dye-sensitized solar cells (Figure 5.10(a)), wherever the dye particles are changed along with SQD (Liu et al., 2011; Sun et al., 2012).

Besides, the theoretical productivity of QDSCs at present is around 44%, the practical performance still trails that of DSSCs. In order to answer the question, Hodes entirely compared the important features of QDSCs, wherefore it is lesser to DSSCs. In QDSCs the key problems are (i) at surface positions quick-moving electron location of the TiO_2/QD intersection attributed toward a leisurely electron inoculation rate through QDs toward TiO_2 (Figure 5.10(b)) and (ii) hole trapping among QDs (Fan et al., 2012; Sun et al., 2012). To attain viable photoconversion productivity with DSSCs, the overhead features in QDSCs have to be bottlenecked. In the background of encouraging QDSC performance, different methods were established—shell layer, doping anode, Pt-free CEs, hierarchical photoanodes, QD passivation, panchromatic sensitizers, etc. (Sun et al., 2012). Nevertheless, for increasing QDSC performance graphene has attracted increasing interest as a possible candidate. He et al. (2011). were the first who established the possibility of QD-decorated graphene (Figure 5.10(c)) as photoelectrode, which displayed the finest IPCE (incident photon to converted electron) related to QD-decorated CNTs scaffold. This approves the graphene as a great candidate for the transport and collection of photogenerated charges, and therefore, the covered nanofilm (Figure 5.10(d)) gives a promising and new direction to evolving high-performance light collecting devices for the subsequent generation solar cells. Following that, another exciting method has been placed, whereas photoanode QD-decorated graphene has been functional in QDSC configuration (Figure 5.10(e) and (f)), ensuing in greater photocurrent than discrete counterparts of QD-based and graphene devices. According to this outcome, the graphene system could assist as a good conducting scaffold to transport and capture photo-induced charge carriers (Anjusree et al., 2013; Chang et al., 2013). Furthermore, the work function of graphene is inferior to the CB of SQD CdSe providing energetic arrangement, which supports the photo persuaded electron proficiently and transfers from excited CdSe to graphene. Thus, an outstanding conductive system of graphene along with high electron transfer rates was attained to spread escape charge recombination and electron mean free paths, safeguarding that by the TCO substrate, electrons can be collected efficiently to yield photocurrent. Analogous phenomena of transporting and capturing electrons from CdS QDs to TCO are grasped in graphene as compared to other carbon nanostructures (He et al., 2011; Kavan et al., 2011a, b).

Although good QD-graphene contact is being shown by QD-decorated a few-layer graphene matrix (Figure 5.10(c) and (d)), the active loading of the QD sensitizer is restricted to just a few monolayers. Hence, to spread

QD films into the 3rd dimension, the careful control of graphene loading is compulsory, where an important balance must lie among maximizing contact with SQD whereas by graphene diminishing incident light absorption, which per monolayer absorbs the 2% of the incident light. Kamat and Lightcap (2012) reveal the innovative architecture of 3D QD-sensitized graphene photoelectrodes, which signifies an important step toward controlling conductivity problems integral to semiconductor quantum dot films and unlocks a new path toward enhanced sensitizer loading. Moreover, this device architecture (Figure 5.10(b)), showed enhanced photocurrent response (which is around 150%) over those ready deprived of GO ascribed to great collection of photogenerated electrons from the GO network in QD-GO compound films, is likely to lessen charge recombination losses at semiconductor quantum dot grain boundaries (Bajpai et al., 2011; Song et al., 2011).

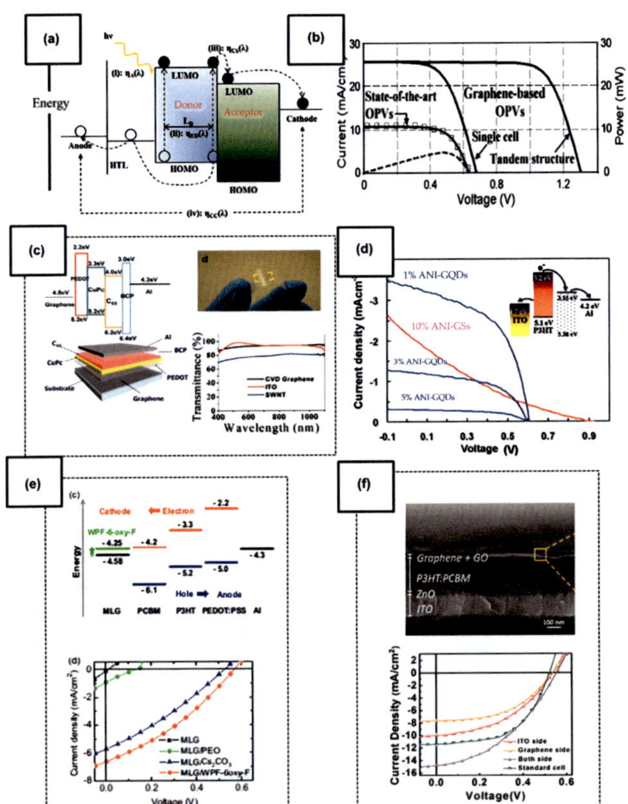

Figure 5.10. (a) The principle of the electron-to-photon alteration procedure in bulk hetero construction solar cells is exposed; (b) the theoretical production

of OPV cells that are graphene-based recommended by Yong and Tour; (c) the representation assemblage of graphene TCE-based displays the model photo and similar transmission spectra along with the OPV band figure of graphene-based OPV; (d) J-V structures based on devices that are photovoltaic aniline-functionalized GQDs with varied GQD contented; (e) energy stage figure of the overturned assembly OPV along with graphene cathode, J, and V features (lowest image); (f) SEM figure of cross-sectional OPV along with graphene cathode layer J and V features of the transparent of solar cells made up of polymer comprising of cathodes that are of graphene (lowermost image).

Source: https://europepmc.org/article/med/21749099.

In mesoscopic solar cells, Song et al. (2011) discovered operationalized GQDs as new sensitizers from the alteration of the bandgap through UV to visible light wavelength. Into graphite, the predictable aggregation of GQDs was overcome from protecting the graphene by enfolding them in all three dimensions from one another. This is attained by covalently joining multiple 1,3,5-trialkyl-substituted phenyl moieties (at 2-position) to the boundaries of graphene (Figure 5.10(c) and (d)). As sensitizers in triiodide-based DSSC conformation the protected GQDs were shown, but it displayed a poor V_{OC} of 0.48 V, and a J_{SC} of 200 1A/cm^2 with a FF of 0.58 were formed (Ding et al., 2011). The extremely lower current density is ascribed to the weak affinity of 1 through the oxide surfaces because of physical adsorption and the resulting weak charge injection from GQDs to TiO_2. Later, the positioning of GQDs (with substrate) was improved as disk-like figures, and it advances the energy adaptation performance (Figure 5.10(e)) as compared to "face-on" positioning because of greater sensitizer filling density that occurs in this structure (Xin et al., 2012). Lately, Williams and Hess (1984) studied at GQD-anchored single crystal TiO_2 (110) the opportunity of hot electron injection; this shows that the graphene nanomaterials have viabilities of applying hot carrier solar cells. However, the CB position of TiO_2 is similar to the HOMO level of GQDs, which permits the collection of electrons [rebound effect; arrow shown in Figure 5.10(f)). Furthermore, the graphene nanostructure has to be made by a dimension of hundreds of nanometers to micrometers, which can be probable by using graphene nanoribbons (GNRs) or graphene flakes. These graphene nanostructures are expected to assist as the hot carrier chromophore permitting multiple electronic excitations below specific experimental conditions. So, it is expected that, at TiO_2/GQD interfaces by the synergetic method of hot electron gathering and diminishing recombination from appropriate interfacial engineering,

GQDs in mesoscopic solar cells are superficial sensitizers (Roy-Mayhew et al., 2010; Kavan et al., 2011c).

5.3.4. Graphene-Based OPVs

OPVs just have gained wide development and research because of their ability as conformal, elastic, lightweight, and cheap power bases for several usages. In Figure 5.10(a) the manual of working of a simple OPV is given, which for photon-to-electron conversion involved in several stages. Primary, an electron in the donor, beneath light absorption experiences photo-induced excitement making a Frenkel excitement, and into the donor-acceptor (D-A) boundaries, these excitons should have to diffuse inside the diffusion distance to avoid collection to the ground state. In terms of charge carriage to a P-N junction in an inorganic semiconductor, this D-A interface concept is similar. Then, at a D-A boundary, an exciton experiences the CT method at a vigorous scale (up to 100 fs) to make a CT exciton, in which the electron and hole persist in the contributor and acceptor points, respectively, keep composed from columbic attraction. As a result of the built-in electric field finally, the CT exciton detaches, into electrons and free holes, then which are passed from the giver and acceptor stages, respectively, toward their particular conductors. OPVs show as a high-scale transformative solar technology and are composed of earth-abundant and nontoxic supplies (Zhang et al., 2010; Guldi and Sgobba, 2011). In large-scale production, in spite of the high capability, their performance could be enhanced for commercial applications. In prototype cells now OPVs display PCEs of 8 to 9%. Theoretical prototypes recommend that the productivity might be enhanced around 10% intended for single-junction cells and 15%for tandem (multi-junction) cells. In order to attain these action goals, developments in the project of constant device assemblies are essential. Furthermore, in determining the act of OPV devices interface engineering also plays a vital part. Graphene-based materials up till now, have been used as transparent electrodes, hole carriers or electron acceptors, etc., in order to elevate the actions of OPV gadgets. Lately, Yong et al. suggested the graphene-based photovoltaics nanostructured can produce greater than 12% (and in a loaded structure it is up to 24%) (Figure 5.0(b)) and it is heavily recommended using functionalized graphene (semiconducting graphene) as the pristine (metallic graphene) and photoactive material in OPV devices as the conductive electrodes (Tang et al., 2010; Wei, 2010).

On graphene a large amount of study work has been presented as transparent conducting electrodes as (Figure 5.0(c)) graphene displays low sheet resistivity ($R_{s,}$ 100 X/sq) in a wide window, high transparency (.80%), and a suitable work function of around 4.5–5.2 eV. At D-A boundaries the separation of charge rate in OPVs is one of the major problems compared to common inorganic solar cells are. In the case of holes and the free electrons, the excitons are produced in the inorganic solar cell's case, but in the case of OPVs, it produces thoroughly bounded hole and electron pairs (Wang et al., 2012; Tang et al., 2013). So, to detach these thoroughly bounded holes and electron pairs, to the fixed at the interface inside the field, an electron acceptor is crucial to breakdown the excitons that diffuse there onto free carriers. At the D-A interfaces for making effective charge separation, at the anode layer, the electron affinity is greater than that of the donor polymer (Yang et al., 2010; Bell et al., 2011). Now, a mixture of the poly(3-octylthiophene) (P3OT) or poly(3-hexylthiophene) (P3HT) with the [6,6]-phenyl-C61-butyric acid methyl ester (PCBM) in OPVs is being used as the electron donor and Acceptor pair (Sun et al., 2010; Neo and Ouyang, 2013). But, considering the quite low PCE of the OPVs for electron acceptors the fullerene-based phenyl-C61-butyric acid methyl ester is not essentially the optimal selection. At D-A interfaces considering the separating electrons, graphene can be regarded as a capable electron acceptor due to their great charge movement of about 7×10^4 $cm^2/V/s$. Furthermore, for a large interface because of their tunable bandgap, high mobility, and high specific surface area, in photovoltaic devices GQDs show high ability as the great electron acceptor material (Figure 5. 10(d)). The photoconversion productivity of cells directs through the layer of the cathode in an OPV device, through which the following properties could be satisfied: (i) the energetic barrier height is being adjusted between the transparent electrodes and the active layer (the V_{OC} of the cell is being determined through it) (Figure 5.10(e)) and (ii) for holes make a selective transporting layer when blocking the electrons. As a potential cathode layer graphene has been commonly functional (Figure 5.10(f)) in the OPVs employing is successfully dealing with the above-mentioned properties. New analyzes by Chen et al. (2009) and Chuchmała and Iwan on the polymer solar cells graphene-based with challenges plus perspectives comprise a thorough literature survey (Chen et al., 2012; Tang and Hu, 2012).

REFERENCES

1. Akturk, A., & Goldsman, N., (2008). Electron transport and full-band electron-phonon interactions in graphene. *Journal of Applied Physics, 103*(5), 053702.
2. Anjusree, G. S., Nair, A. S., Nair, S. V., & Vadukumpully, S., (2013). One-pot hydrothermal synthesis of TiO_2/graphene nanocomposites for enhanced visible-light photocatalysis and photovoltaics. *RSC Advances, 3*(31), 12933–12938.
3. Bae, S., Kim, H., Lee, Y., Xu, X., Park, J. S., Zheng, Y., & Kim, Y. J., (2010). Roll-to-roll production of 30-inch graphene films for transparent electrodes. *Nature Nanotechnology, 5*(8), 574.
4. Bajpai, R., Roy, S., Kumar, P., Bajpai, P., Kulshrestha, N., Rafiee, J., & Misra, D. S., (2011). Graphene supported platinum nanoparticle counter-electrode for enhanced performance of dye-sensitized solar cells. *ACS Applied Materials and Interfaces, 3*(10), 3884–3889.
5. Balandin, A. A., (2011). Thermal properties of graphene and nanostructured carbon materials. *Nature Materials, 10*(8), 569–581.
6. Becerril, H. A., Mao, J., Liu, Z., Stoltenberg, R. M., Bao, Z., & Chen, Y., (2008). Evaluation of solution-processed reduced graphene oxide films as transparent conductors. *ACS Nano, 2*, 463–470.
7. Bell, N. J., Ng, Y. H., Du, A., Coster, H., Smith, S. C., & Amal, R., (2011). Understanding the enhancement in photoelectrochemical properties of photocatalytically prepared TiO_2-reduced graphene oxide composite. *The Journal of Physical Chemistry C, 115*(13), 6004–6009.
8. Berger, C., Song, Z., Li, T., Li, X., Ogbazghi, A. Y., Feng, R., & De Heer, W. A., (2004). Ultrathin epitaxial graphite: 2D electron gas properties and a route toward graphene-based nanoelectronics. *The Journal of Physical Chemistry B, 108*(52), 19912–19916.
9. Chang, J., Yang, J., Ma, P., Wu, D., Tian, L., Gao, Z., & Yang, L., (2013). Hierarchical titania mesoporous sphere/graphene composite, synthesis, and application as photoanode in dye-sensitized solar cells. *Journal of Colloid and Interface Science, 394*, 231–236.
10. Chavez-Valdez, A. R. B. A., Shaffer, M. S., & Boccaccini, A. R., (2013). Applications of graphene electrophoretic deposition. A review. *The Journal of Physical Chemistry B,117*(6), 1502–1515.
11. Chen, C., Yang, Q. H., Yang, Y., Lv, W., Wen, Y., Hou, P. X., & Cheng, H. M., (2009). Self-assembled free-standing graphite oxide membrane.

Advanced Materials, 21(29), 3007–3011.

12. Chen, D., Zhang, H., Liu, Y., & Li, J., (2013). Graphene and its derivatives for the development of solar cells, photoelectrochemical, and photocatalytic applications. *Energy and Environmental Science, 6*(5), 1362–1387.
13. Chen, J. H., Jang, C., Xiao, S., Ishigami, M., & Fuhrer, M. S., (2008). Intrinsic and extrinsic performance limits of graphene devices on SiO_2. *Nature Nanotechnology, 3*(4), 206–209.
14. Chen, S. Q., & Wang, Y., (2010). Microwave-assisted synthesis of a Co_3O_4-graphene sheet-on-sheet nanocomposite as a superior anode material for Li-ion batteries. *Journal of Materials Chemistry, 20*(43), 9735–9739.
15. Chen, T., Hu, W., Song, J., Guai, G. H., & Li, C. M., (2012). Interface functionalization of photoelectrodes with graphene for high performance dye-sensitized solar cells. *Advanced Functional Materials, 22*(24), 5245–5250.
16. Choi, W., & Lee, J. W., (2011). *Graphene: Synthesis and Applications* (Vol. 1, pp. 1–33). CRC press.
17. Cote, L. J., Kim, F., & Huang, J., (2009). Langmuir-Blodgett assembly of graphite oxide single layers. *Journal of the American Chemical Society, 131*(3), 1043–1049.
18. Das, S., Sudhagar, P., Verma, V., Song, D., Ito, E., Lee, S. Y., & Choi, W., (2011). Amplifying charge-transfer characteristics of graphene for triiodide reduction in dye-sensitized solar cells. *Advanced Functional Materials, 21*(19), 3729–3736.
19. De Heer, W. A., Berger, C., Wu, X., First, P. N., Conrad, E. H., Li, X., & Potemski, M., (2007). Epitaxial graphene. *Solid State Communications, 143*(1, 2), 92–100.
20. Ding, J. N., Yu, C. T., Yuan, N. Y., Liu, Y. B., & Fan, Y., (2011). High-quality GS/TiO_2 composite for the photoanode of the dye-sensitized solar cells. In: *2011 International Conference on Materials for Renewable Energy and Environment* (Vol. 1, pp. 90–94). IEEE.
21. Eda, G., Lin, Y. Y., Miller, S., Chen, C. W., Su, W. F., & Chhowalla, M., (2008). Transparent and conducting electrodes for organic electronics from reduced graphene oxide. *Applied Physics Letters, 92*(23), 209.
22. Emtsev, K. V., Bostwick, A., Horn, K., Jobst, J., Kellogg, G. L., Ley, L., & Rotenberg, E., (2009). Towards wafer-size graphene layers

by atmospheric pressure graphitization of silicon carbide. *Nature Materials, 8*(3), 203–207.

23. Fan, J., Liu, S., & Yu, J., (2012). Enhanced photovoltaic performance of dye-sensitized solar cells based on TiO_2 nanosheets/graphene composite films. *Journal of Materials Chemistry, 22*(33), 17027–17036.
24. Geim, A. K., & Novoselov, K. S., (2010). The rise of graphene. In: *Nanoscience and Technology: A Collection of Reviews from Nature Journals* (pp. 11–19).
25. Gong, F., Xu, X., Zhou, G., & Wang, Z. S., (2013). Enhanced charge transportation in a polypyrrole counter electrode via incorporation of reduced graphene oxide sheets for dye-sensitized solar cells. *Physical Chemistry Chemical Physics, 15*(2), 546–552.
26. Guldi, D. M., & Sgobba, V., (2011). Carbon nanostructures for solar energy conversion schemes. *Chemical Communications, 47*(2), 606–610.
27. Hagfeldt, A., Boschloo, G., Sun, L., Kloo, L., & Pettersson, H., (2010). Dye-sensitized solar cells. *Chemical Reviews, 110*(11), 6595–6663.
28. Hasin, P., Alpuche-Aviles, M. A., & Wu, Y., (2010). Electrocatalytic activity of graphene multilayers toward I−/I3−: Effect of preparation conditions and polyelectrolyte modification. *The Journal of Physical Chemistry C,114*(37), 15857–15861.
29. He, Z., Guai, G., Liu, J., Guo, C., Loo, J. S. C., Li, C. M., & Tan, T. T. Y., (2011). Nanostructure control of graphene-composited TiO_2 by a one-step solvothermal approach for high performance dye-sensitized solar cells. *Nanoscale, 3*(11), 4613–4616.
30. Hernandez, Y., Nicolosi, V., Lotya, M., Blighe, F. M., Sun, Z., De, S., & Boland, J. J., (2008). High-yield production of graphene by liquid-phase exfoliation of graphite. *Nature Nanotechnology, 3*(9), 563–568.
31. Hummers, Jr. W. S., & Offeman, R. E., (1958). Preparation of graphitic oxide. *Journal of the American Chemical Society, 80*(6), 1339–1339.
32. Hupalo, M., Conrad, E. H., & Tringides, M. C., (2009). Growth mechanism for epitaxial graphene on vicinal 6 H-SiC (0001) surfaces: A scanning XXXunnelling microscopy study. *Physical Review B,80*(4), 041401.
33. Ishikawa, R., Bando, M., Wada, H., Kurokawa, Y., Sandhu, A., & Konagai, M., (2012). Layer-by-layer assembled transparent conductive

graphene films for silicon thin-film solar cells. *Japanese Journal of Applied Physics, 51*(11S), 11PF01.

34. Kavan, L., Yum, J. H., & Grätzel, M., (2011a). Graphene nanoplatelets outperforming platinum as the electrocatalyst in co-bipyridine-mediated dye-sensitized solar cells. *Nano Letters, 11*(12), 5501–5506.
35. Kavan, L., Yum, J. H., & Grätzel, M., (2011c). Optically transparent cathode for dye-sensitized solar cells based on graphene nanoplatelets. *ACS Nano, 5*(1), 165–172.
36. Kavan, L., Yum, J. H., Nazeeruddin, M. K., & Grätzel, M., (2011b). Graphene nanoplatelet cathode for Co (III)/(II) mediated dye-sensitized solar cells. *ACS Nano, 5*(11), 9171–9178.
37. Kim, C. H., Kim, B. H., & Yang, K. S., (2012). TiO_2 nanoparticles loaded on graphene/carbon composite nanofibers by electrospinning for increased photocatalysis. *Carbon, 50*(7), 2472–2481.
38. Kim, H. N., Yoo, H., & Moon, J. H., (2013). Graphene-embedded 3D TiO_2 inverse opal electrodes for highly efficient dye-sensitized solar cells: Morphological characteristics and photocurrent enhancement. *Nanoscale, 5*(10), 4200–4204.
39. Kim, J. Y., Lee, K., Coates, N. E., Moses, D., Nguyen, T. Q., Dante, M., & Heeger, A. J., (2007). Efficient tandem polymer solar cells fabricated by all-solution processing. *Science, 317*(5835), 222–225.
40. Kim, J., Cote, L. J., Kim, F., Yuan, W., Shull, K. R., & Huang, J., (2010). Graphene oxide sheets at interfaces. *Journal of the American Chemical Society, 132*(23), 8180–8186.
41. Kymakis, E., Stratakis, E., Stylianakis, M. M., Koudoumas, E., & Fotakis, C., (2011). Spin coated graphene films as the transparent electrode in organic photovoltaic devices. *Thin Solid Films, 520*(4), 1238–1241.
42. Lahiri, I., Verma, V. P., & Choi, W., (2011). An all-graphene based transparent and flexible field emission device. *Carbon, 49*(5), 1614–1619.
43. Lee, J. S., Ahn, H. J., Yoon, J. C., & Jang, J. H., (2012a). Three-dimensional nano-foam of few-layer graphene grown by CVD for DSSC. *Physical Chemistry Chemical Physics, 14*(22), 7938–7943.
44. Lee, K. S., Lee, Y., Lee, J. Y., Ahn, J. H., & Park, J. H., (2012b). Flexible and platinum-free dye-sensitized solar cells with conducting-polymer-coated graphene counter electrodes. *ChemSusChem, 5*(2), 379–382.

45. Li, D., Müller, M. B., Gilje, S., Kaner, R. B., & Wallace, G. G., (2008). Processable aqueous dispersions of graphene nanosheets. *Nature Nanotechnology, 3*(2), 101–105.
46. Li, G. L., Liu, G., Li, M., Wan, D., Neoh, K. G., & Kang, E. T., (2010). Organo-and water-dispersible graphene oxide-polymer nanosheets for organic electronic memory and gold nanocomposites. *The Journal of Physical Chemistry C, 114*(29), 12742–12748.
47. Li, S. S., Tu, K. H., Lin, C. C., Chen, C. W., & Chhowalla, M., (2010). Solution-processable graphene oxide as an efficient hole transport layer in polymer solar cells. *ACS Nano, 4*(6), 3169–3174.
48. Li, X., Cai, W., An, J., Kim, S., Nah, J., Yang, D., & Banerjee, S. K., (2009a). Large-area synthesis of high-quality and uniform graphene films on copper foils. *Science, 324*(5932), 1312–1314.
49. Li, X., Zhang, R., Yu, W., Wang, K., Wei, J., Wu, D., & Ruoff, R. S., (2012). Stretchable and highly sensitive graphene-on-polymer strain sensors. *Scientific Reports, 2*(870), 1–6.
50. Li, X., Zhu, H., Wang, K., Cao, A., Wei, J., Li, C., & Wu, D., (2010). Graphene-on-silicon Schottky junction solar cells. *Advanced Materials, 22*(25), 2743–2748.
51. Li, X., Zhu, Y., Cai, W., Borysiak, M., Han, B., Chen, D., & Ruoff, R. S., (2009b). Transfer of large-area graphene films for high-performance transparent conductive electrodes. *Nano Letters, 9*(12), 4359–4363.
52. Lightcap, I. V., & Kamat, P. V., (2012). Fortification of CdSe quantum dots with graphene oxide. Excited state interactions and light energy conversion. *Journal of the American Chemical Society, 134*(16), 7109–7116.
53. Liu, X., Pan, L., Lv, T., Zhu, G., Sun, Z., & Sun, C., (2011). Microwave-assisted synthesis of CdS-reduced graphene oxide composites for photocatalytic reduction of Cr (vi). *Chemical Communications, 47*(43), 11984–11986.
54. Madhavan, A. A., Kalluri, S., Chacko, D. K., Arun, T. A., Nagarajan, S., Subramanian, K. R., & Balakrishnan, A., (2012). Electrical and optical properties of electrospun TiO_2-graphene composite nanofibers and its application as DSSC photo-anodes. *RSC Advances, 2*(33), 13032–13037.
55. Miao, X., Tongay, S., Petterson, M. K., Berke, K., Rinzler, A. G., Appleton, B. R., & Hebard, A. F., (2012). High efficiency graphene

solar cells by chemical doping. *Nano Letters, 12*(6), 2745–2750.

56. Morales-Torres, S., Pastrana-Martínez, L. M., Figueiredo, J. L., Faria, J. L., & Silva, A. M., (2012). Design of graphene-based TiO_2 photocatalysts: A review. *Environmental Science and Pollution Research, 19*(9), 3676–3687.
57. Mora-Seró, I., & Bisquert, J., (2010). Breakthroughs in the development of semiconductor-sensitized solar cells. *The Journal of Physical Chemistry Letters, 1*(20), 3046–3052.
58. Neo, C. Y., & Ouyang, J., (2013). Graphene oxide as auxiliary binder for TiO_2 nanoparticle coating to more effectively fabricate dye-sensitized solar cells. *Journal of Power Sources, 222*, 161–168.
59. Novoselov, K. S., Geim, A. K., Morozov, S. V., Jiang, D., Zhang, Y., Dubonos, S. V., & Firsov, A. A., (2004). Electric field effect in atomically thin carbon films. *Science, 306*(5696), 666–669.
60. Roy-Mayhew, J. D., Bozym, D. J., Punckt, C., & Aksay, I. A., (2010). Functionalized graphene as a catalytic counter electrode in dye-sensitized solar cells. *ACS Nano, 4*(10), 6203–6211.
61. Song, J., Yin, Z., Yang, Z., Amaladass, P., Wu, S., Ye, J., & Liu, X. W., (2011). Enhancement of photogenerated electron transport in dye-sensitized solar cells with introduction of a reduced graphene oxide-TiO_2 junction. *Chemistry: A European Journal, 17*(39), 10832–10837.
62. Sun, L., Zhao, Z., Zhou, Y., & Liu, L., (2012). Anatase TiO_2 nanocrystals with exposed {001} facets on graphene sheets via molecular grafting for enhanced photocatalytic activity. *Nanoscale, 4*(2), 613–620.
63. Sun, S., Gao, L., & Liu, Y., (2010). Enhanced dye-sensitized solar cell using graphene-TiO_2 photoanode prepared by heterogeneous coagulation. *Applied Physics Letters, 96*(8), 083113.
64. Tang, B., & Hu, G., (2012). Two kinds of graphene-based composites for photoanode applying in dye-sensitized solar cell. *Journal of Power Sources, 220*, 95–102.
65. Tang, B., Hu, G., Gao, H., & Shi, Z., (2013). Three-dimensional graphene network-assisted high performance dye-sensitized solar cells. *Journal of Power Sources, 234*, 60–68.
66. Tang, Y. B., Lee, C. S., Xu, J., Liu, Z. T., Chen, Z. H., He, Z., & Luo, L., (2010). Incorporation of graphenes in nanostructured TiO_2 films via molecular grafting for dye-sensitized solar cell application. *ACS Nano, 4*(6), 3482–3488.

67. Tongay, S., Berke, K., Lemaitre, M., Nasrollahi, Z., Tanner, D. B., Hebard, A. F., & Appleton, B. R., (2011). Stable hole doping of graphene for low electrical resistance and high optical transparency. *Nanotechnology, 22*(42), 425701.
68. Tsai, T. H., Chiou, S. C., & Chen, S. M., (2011). Enhancement of dye-sensitized solar cells by using graphene-TiO_2 composites as photoelectrochemical working electrode. *Int. J. Electrochem. Sci., 6*(8), 3333–3343.
69. Tu, W., Zhou, Y., & Zou, Z., (2013). Versatile graphene-promoting photocatalytic performance of semiconductors: Basic principles, synthesis, solar energy conversion, and environmental applications. *Advanced Functional Materials, 23*(40), 4996–5008.
70. Verma, V. P., Das, S., Lahiri, I., & Choi, W., (2010). Large-area graphene on polymer film for flexible and transparent anode in field emission device. *Applied Physics Letters, 96*(20), 203108.
71. Viculis, L. M., Mack, J. J., & Kaner, R. B., (2003). A chemical route to carbon nanoscrolls. (Brevia). *Science, 299*(5611), 1361–1362.
72. Wang, D. H., Kim, J. K., Seo, J. H., Park, I., Hong, B. H., Park, J. H., & Heeger, A. J., (2013). Transferable graphene oxide by stamping nanotechnology: electron-transport layer for efficient bulk-heterojunction solar cells. *Angewandte Chemie International Edition, 52*(10), 2874–2880.
73. Wang, H. X., Wang, Q., Zhou, K. G., & Zhang, H. L., (2013). Graphene is light: Design, synthesis and applications of photo-active graphene and graphene-like materials. *Small, 9*(8), 1266–1283.
74. Wang, H., Leonard, S. L., & Hu, Y. H., (2012). Promoting effect of graphene on dye-sensitized solar cells. *Industrial and Engineering Chemistry Research, 51*(32), 10613–10620.
75. Wang, X., Zhi, L., & Müllen, K., (2008). Transparent, conductive graphene electrodes for dye-sensitized solar cells. *Nano Letters, 8*(1), 323–327.
76. Wassei, J. K., Mecklenburg, M., Torres, J. A., Fowler, J. D., Regan, B. C., Kaner, R. B., & Weiller, B. H., (2012). Chemical vapor deposition of graphene on copper from methane, ethane, and propane: Evidence for bilayer selectivity. *Small, 8*(9), 1415–1422.
77. Wei, D., (2010). Dye-sensitized solar cells. *International Journal of Molecular Sciences, 11*(3), 1103–1113.

78. Williams, L. M., & Hess, D. W., (1984). Photoelectrochemical properties of plasma-deposited TiO_2 thin films. *Thin Solid Films, 115*(1), 13–18.

79. Wintterlin, J., & Bocquet, M. L., (2009). Graphene on metal surfaces. *Surface Science, 603*(10–12), 1841–1852.

80. Xin, X., Zhou, X., Wu, J., Yao, X., & Liu, Z., (2012). Scalable synthesis of TiO_2/graphene nanostructured composite with high-rate performance for lithium-ion batteries. *ACS Nano, 6*(12), 11035–11043.

81. Yang, N., Zhai, J., Wang, D., Chen, Y., & Jiang, L., (2010). Two-dimensional graphene bridges enhanced photoinduced charge transport in dye-sensitized solar cells. *ACS Nano, 4*(2), 887–894.

82. Ye, Y., Dai, Y., Dai, L., Shi, Z., Liu, N., Wang, F., & Liu, Z., (2010). High-performance single CdS nanowire (nanobelt) Schottky junction solar cells with Au/graphene Schottky electrodes. *ACS Applied Materials and Interfaces, 2*(12), 3406–3410.

83. Yella, A., Lee, H. W., Tsao, H. N., Yi, C., Chandiran, A. K., Nazeeruddin, M. K., & Grätzel, M., (2011). Porphyrin-sensitized solar cells with cobalt (II/III)-based redox electrolyte exceed 12 percent efficiency. *Science, 334*(6056), 629–634.

84. Zhang, J., Zhao, F., Zhang, Z., Chen, N., & Qu, L., (2013). Dimension-tailored functional graphene structures for energy conversion and storage. *Nanoscale, 5*(8), 3112–3126.

85. Zhang, X. Y., Li, H. P., Cui, X. L., & Lin, Y., (2010). Graphene/TiO_2 nanocomposites: Synthesis, characterization, and application in hydrogen evolution from water photocatalytic splitting. *Journal of Materials Chemistry, 20*(14), 2801–2806.

86. Zhang, X., Suresh, K. P., Aravindan, V., Liu, H. H., Sundaramurthy, J., Mhaisalkar, S. G., & Madhavi, S., (2012). Electrospun TiO_2-graphene composite nanofibers as a highly durable insertion anode for lithium-ion batteries. *The Journal of Physical Chemistry C,116*(28), 14780–14788.

87. Zhu, P., Nair, A. S., Shengjie, P., Shengyuan, Y., & Ramakrishna, S., (2012). Facile fabrication of TiO_2-graphene composite with enhanced photovoltaic and photocatalytic properties by electrospinning. *ACS Applied Materials and Interfaces, 4*(2), 581–585.

88. Zhu, Y., Cai, W., Piner, R. D., Velamakanni, A., & Ruoff, R. S., (2009). Transparent self-assembled films of reduced graphene oxide platelets. *Applied Physics Letters, 95*(10), 103104.

CHAPTER 6

Introduction to Graphene Photonics

CONTENTS

6.1. INTRODUCTION

As is the case with electronics, the future and present of photonics appear to be declining, specifically, in the size of the device. In this pattern, 2D graphene is exhibiting future potential in these systems and devices. For high-speed optoelectronic devices, the high-carrier mobility of graphene is an advantage. Uses in a range of spectral areas will gain from broad spectral absorption of graphene (Yamashita, 2011; Wang et al., 2013). For applications needing nonlinear optical structures and robust photonic high optical harm threshold and the mechanical strength of graphene will be highest. Additionally, the simplicity of transfer of graphene to some other substrates and the ability of graphene to conformally cover surfaces make graphene perfect for incorporating hybrid photonic systems (Wallace, 1947; Kamisan et al., 2015).

This chapter intends to give a fundamental understanding of the uses of graphene in photonic devices. The chapter is split into five parts:

- Photo-detection;
- Optical polarization;
- Electro-optic modulation;
- Plasmonics; and
- Nonlinear devices and optical properties.

6.2. PHOTODETECTORS

Three chief classes of photodetectors have normally been made using graphene. In the 1st group, photons generally couple to electrons generating an electronic response, this response defines the photovoltaic style detectors. The 2nd category includes coupling photons, phonons, and electrons and outcomes in the hot-electron bolometers. Lastly, the 3rd embodiment uses the amalgam photodetector of graphene: one of the materials absorbs light and is incorporated onto graphene in order to exploit transport properties and carrier collection of graphene (Bao and Loh, 2012; Avouris and Freitag, 2013).

In Figure 6.1 some photodetectors of graphene discoursed in this segment are compared with one another along with the photodetectors available commercially. Responsivity is employed as the figure of standard for contrast with telecommunications. Conversely, responsivity is the function of wavelength:

$R = \eta\lambda/1240$ A/W

[λ in nm]

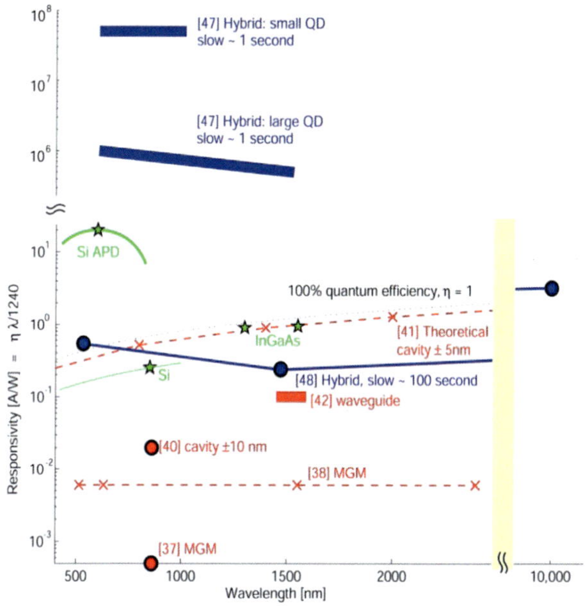

Figure 6.1. The spectral responsivity of photodetectors of graphene matched with the photodetectors available commercially.

Note: The dotted line displays 100% quantum effectiveness, η=1, as the function of wavelength. Green and red colors represent ≤1-nano seconds response time, whereas the blue color represents ≥one-sec response times. The photodetectors of graphene are marked with the reference number along with the short description of the style of photodetector: planar MGM, waveguide device, optical resonant cavity, and the mix graphene photodetector. The photodiodes available commercially, exhibited in green color, are the Hamamatsu models S8664 (Si APD-avalanche photodiode), S9055 (Si), and G8195 (InGaAs).

Therefore, the dotted line representing a 100% quantum efficiency adds the reference for the spectral comparison. Rapid photodetectors of graphene have responses ≤ one nanoseconds and are represented with the red color, whereas the one in blue color denotes a slow response time of ≥ one second. The photodetectors available commercially (Si APD, InGaAs, and Si,) all have rapid responses ≤ 1 nanosecond (Figures 6.2–6.4) (Moon and Gaskill, 2011; Sun et al., 2013).

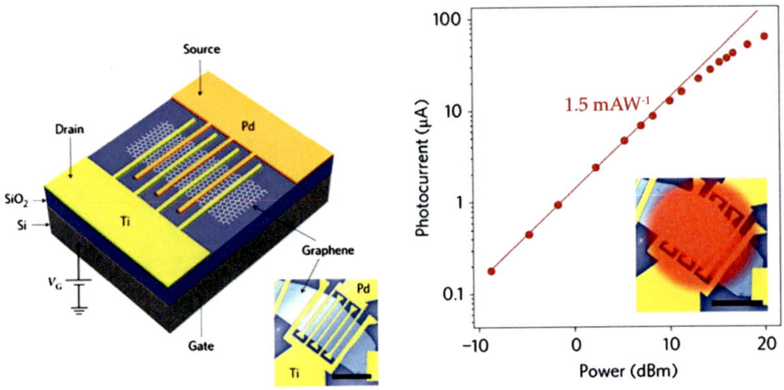

Figure 6.2. Drawing of metal-graphene-metal photodetector and its fabricated structure as well as the photocurrent response at the wavelength of 1550 nanometer(Moon and Gaskill, 2011; Sun et al., 2013).

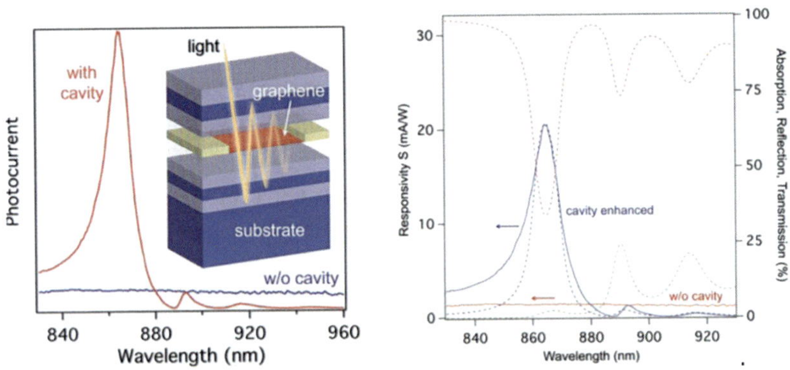

Figure 6.3. The photodetector with graphene microcavity traps light utilizing disseminated Bragg mirrors in the multi-pass cavity comprising graphene. The cavity generally increases the absorption of the single layer of graphene, thus escalating responsivity at the cost of the spectral bandwidth of the cavity.

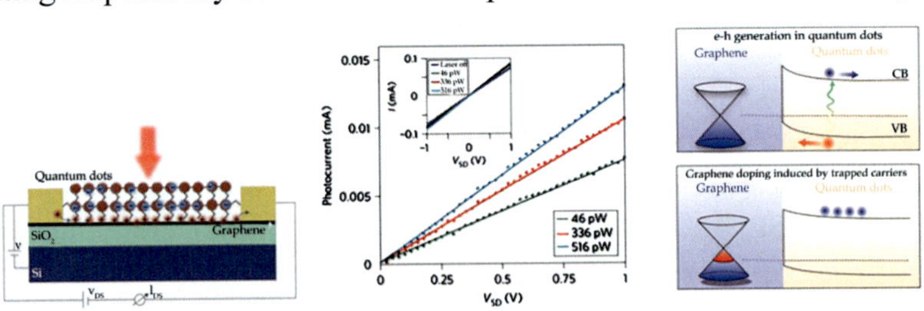

Figure 6.4. Amalgam photodetector utilizing quantum dots of PbS as the photodetection transduce varying incident light into the mobile carriers. As displayed

in the diagram of energy-band, electrons are usually stuck in the dots, whereas holes are transported to the graphene and yield a photocurrent.

There are three main types of photodetectors based on graphene:

- Photothermal detectors;
- Photovoltaic detectors; and
- Hybrid-graphene photodetectors.

6.3. ELECTRO-OPTIC MODULATION

Light modulation is required in several photonic applications. The materials can be utilized for electro-optic light modulation if some of the characteristics of light (polarization, amplitude, etc.), can usually be altered by an injected current or applied electric field. Electro-optic light modulation is normally accomplished by various popular methods, like:

- The electro-optic material;
- The electro-absorption (EA) material;
- electrically tempted modulation of the refractive index of the optical material;
- Electrically tempted thermal discrepancies which alter the properties of an optical material.

An EA modulator based on graphene was demonstrated experimentally by depositing the monolayer of graphene on the optical waveguide (Geim and Grigorieva, 2013). The highly-restricted optical field in the SiO/Si waveguide intermingled with the oxide cap and graphene through the passing field. An electric field has usually been displayed to alter the electron-phonon coupling, conductivity, Fermi level, and work function in graphene. Chou et al. (1992)utilized an electric field in order to adjust the Fermi level of the sheet of graphene, thus, altering the optical absorption/conductivity and making an optical modulator (Grigorenko et al., 2012). With the help of this approach, they attained an EA modulation of around 0.1 dB/μm, validating the device with a small footprint (around 25 μm^2). Almost greater than one-Giga Hz modulation frequency was attained over the wide optical bandwidth (1.351.6 μm). Furthermore, the modulation of light was polarization-sensitive as the graphene intermingles with the electric field (in-plane) of the passing optical wave (Wang et al., 2012).

EA modulators are usual in optical telecommunications as these modulators often function with a lower voltage bias as compared to the modulators created with the electro-optic materials. The function of EA modulation is frequently accomplished by shifting the absorption energies, normally in quantum-restricted structures and semiconductors by employing the Quantum-confined Stark or Franz-Keldysh effects. One disadvantage of EA modulators is that the optical power absorbed can restrict device operation. Conversely, the high thermal conductivity of graphene might overpower this potential restriction by quickly moving the heat away (Figure 6.5) (Adachi, 1985; Chen et al., 2008).

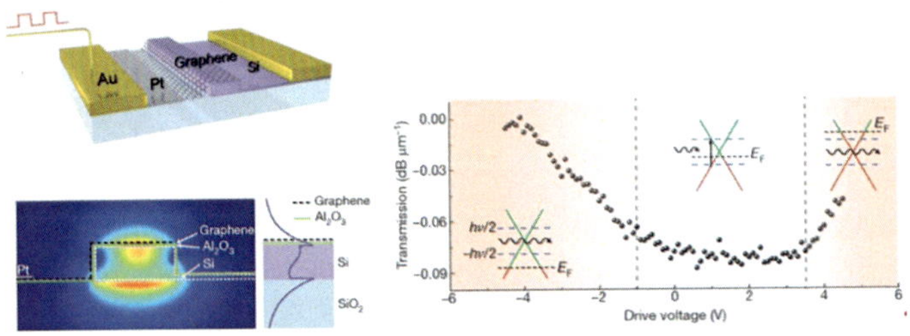

Figure 6.5. Graphic representation and the cross-section of the optical mode in an electro-absorption modulator based on graphene along with the plot of E-field modulation of the optical transmission with diagrams of energy-band enlightening the field-tempted change in fermi level.

In one other current experiment, graphene's optical transmission was modulated with the help of coplanar electrodes on the multilayer graphene. The simple experiment hardly placed a voltage amongst the electrodes and then measured a small variation in the optical transmission over the wide optical bandwidth. The modulation was quite slow and permits higher harmonics of modulation frequency through. Additional work is required to comprehend the physics of the effect; conversely, it is a simple idea that it might be broadly applicable (Burke et al., 2000).

Another approach for modulation of light was currently demonstrated using reflection in THz and the mid-infrared region. By smearing a magnetic field on the multi-layer structures of graphene, the inter-layer coupling was examined by the magneto-optic Kerr spectroscopy which is polarization-sensitive. The experiments perceived cyclotron resonant Kerr characteristics that validated optical reflection persuaded by the external magnetic field. The understanding obtained from the experiment recommends the potential for substituting the external magnetic field with the electrostatic gate (Zhang et al., 2005; Pan et al., 2011).

All these discoveries in the electrically-tempted modulation of light by graphene have occasioned within the last few years. This might be the start for unlocking the prospective of electrically-tempted modulation of light by graphene (Nair et al., 2008).

6.4. POLARIZERS

Applying graphene on the top of optical waveguides permits the optical field to momentarily couple to graphene. It was mentioned formerly in the waveguide-based EA modulator of graphene and in the waveguide-based photodetectors of graphene. The same approach is used to make an optical polarizer. The polarization comes from the graphene collaboration with the electric field (in-plane) of the passing optical wave. This approach was validated in the optical fiber with single-mode, which had been refined to the core of fiber, making a flat region where the graphene was moved (Currie et al., 2011). The distance of propagation, L_G, was stated as the length covered by the film of graphene. At the wavelength of around 1550 nanometer, increasing L_G from 2–7 mm occasioned extinction ratios of polarization to increase from 22–27 dB. The measured extinction of polarization from the monolayer graphene was almost the same as when the few-layer graphene was gathered on the fiber. Polarization over the wide spectral bandwidth from 400–2000 nm was attained. TM and TE coupling and absorption by graphene were discoursed as the polarization mechanism of a waveguide, with the maximum polarization extinction ratio of nearly 27 dB (Figure 6.6) (Bonse et al., 2002).

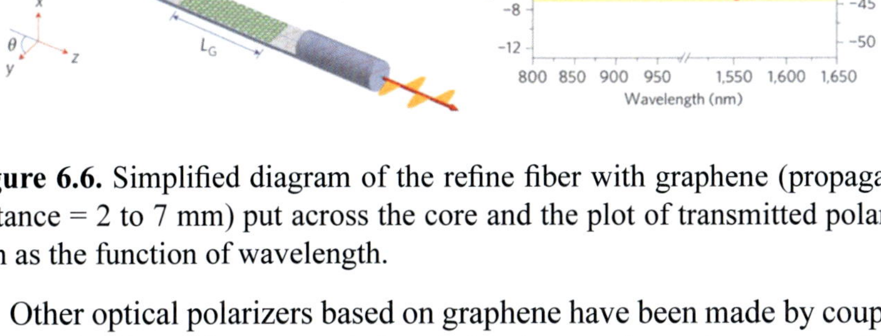

Figure 6.6. Simplified diagram of the refine fiber with graphene (propagation distance = 2 to 7 mm) put across the core and the plot of transmitted polarization as the function of wavelength.

Other optical polarizers based on graphene have been made by coupling to SPP (surface plasmon polaritons) in the graphene. In one particular case, attenuated complete reflectance geometry was well-thought-out with graphene at the total internal reflective surface (Xing et al., 2010). When the condition of resonance is met amongst the incident EM (electromagnetic) wave and the polaritons mode in graphene, then the SPPs are excited in the graphene, and reflectivity of the EM wave is decreased. The coupling of TM and TE polarization are feasible dependent on the imaginary part of intricate conductivity for the efficient dielectric material neighboring the graphene. Additionally, the position of Fermi level in the graphene can impact this coupling, so adding the voltage bias to graphene might permit alteration of the polarizer. The extent of polarization can be controlled by changing the voltage applied. The authors comment on the utilization of this for the tunable polarizer in the frequency range of THz, conversely, it might also be feasible to utilize this as the THz modulator. Ultimately, as this is the theoretical study it is possibly appropriate for optical frequencies with suitable dielectric materials (Berger et al., 2004; Nyakiti, 2012).

6.5. PLASMONICS

These possess the aptitude to couple light into the nanostructures that are usually smaller than 1/ten of the light wavelength. This is very striking as it gives the chance of matter and light interactions on the deep subwavelength scale. For photonics, this permits integration of electronics and light, enabling the plasmonic lenses, concentrating on nanoscale, and the engineered metamaterials. Additionally, the confinement permits EM field improvements and can give unusual nonlinear electronic and optic phenomena (VanMil et al., 2009).

Plasmas are several-body carrier effects which ascend through resonant mutual oscillations of the lightly damped carriers. These carrier effects can happen in the free-electron systems: metals, gases, and doped semiconductors. In graphene, the quantized plasma modes are restricted to the surface and the interface amongst the graphene and neighboring dielectric, therefore, surface plasmons are well-thought-out. The light coupling with the two-dimensional (2D) carriers can be defined by SPP. Dependent on the material, the light varies from THz to Ultraviolet wavelengths, thus this is striking for several areas of photonics comprising imaging, sensing, and communication (Kim et al., 2014; Bae et al., 2010).

Graphene has exclusive properties that make it a valuable plasmonic material. As the semi-metal, the carrier concentration of graphene can be varied. This gives a chance for modifying the plasmon resonance. For instance, the gate voltage could be applied to modify the carrier concentration, thus altering the plasmonic behavior (Lee et al., 2008).

Additionally, carriers of graphene are restricted to 2-dimensions, tough, this 2D system survives in the three-dimensional (3D) world, and thus neighboring dielectric materials impact its response. This can be utilized in the gate voltage in order to modify the resonance of the plasmon. Additionally, epitaxial graphene on the SiC displays radically different behavior as compared to that of the suspended graphene. This behavior is described by the graphene's high doping caused by the substrate of SiC which impacts the coupling amongst the plasmon of graphene and surface optical phonon of SiC (Bernardi et al., 2013).

The Dirac fermions without mass in graphene have the linear electronic dispersion, but the plasmons of graphene possess the nonlinear dispersion relationship. Plasmons having long-wavelength in the Dirac system have the frequency proportional to in all of the dimensions (one-D, two-D, or three-D). This is the non-classical outcome and varies radically from the plasmon frequency with long-wavelength in the classical 2-D carrier system. Furthermore, the frequency of plasmon in graphene shows a 4th root carrier density dependency ($\omega_0 \propto {}^4$), as contrasting to square root dependency in the other 2-D electronic systems (Currie et al., 2013). In addition to their dissimilarities, there exist similarities to the classical plasmons. One instance is that the long-wavelength of graphene plasmon dispersion is directly proportional to the square root of wave vector ($\omega = \omega_0$), as exhibited in Figure 6.7 where at the low-energies the dispersion equals the quadratic dependency of the Drude model (displayed as dashed lines).

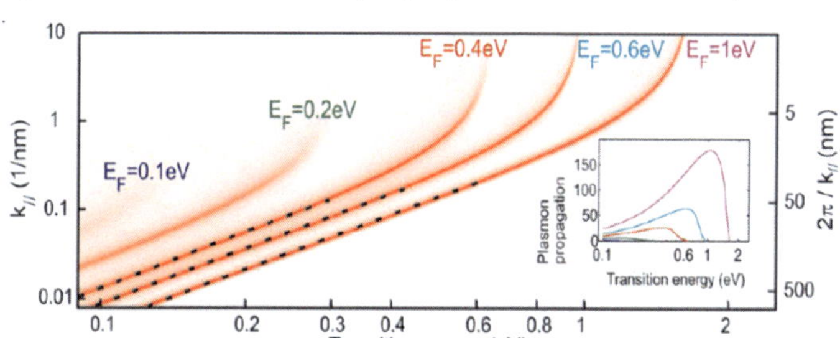

Figure 6.7. The contour plot demonstrates the dispersion of plasmon in the doped graphene. The contours display the magnitude of the Fresnel reflection coefficient as the function of a Fermi level. These illustrate the quadratic dependency on the energy of plasmon with dashed lines conforming to the Drude model. The inset displays the distance of propagation.

Dissimilar to plasmons in the metals which inclines to be quite lossy with resonances restricted to the UV and visible, properties of graphene permit lower losses and function in the THz and infrared (Koh et al., 2013). This is because of the tight confinement of graphene of the SPP directed mode. This characteristic permits the large ratio of plasmon's wavelength to the wavelength of free-space and allows long plasmon lengths of propagation. For instance, the 10-μm free-space wavelength can usually couple to the graphene SPP of almost 144-nm wavelength, signifying a 69x contraction in the size of free-space optical mode to the SPP-directed mode. Also, the length of propagating of this directed SPP, described by the imaginary part of SPP wavenumber, is nearly 2.25 μm, which signifies 15.6x the wavelength of SPP (Ito et al., 2004). This length is quite long to make plasmonic devices, whereas also confirming isolation between neighboring devices.

Centered upon the basic properties, a range of uses have been developed using graphene plasmonics, like optical polarizers, photodetectors, and optical modulators. This was mentioned previously in the circumstance of the tunable polarizer, where the EM waves coupled to the SPP of graphene (Gupta et al., 1991). Plasmonic resonances in mid-IR, 900 to 2500 cm^{-1} was perceived by making nanostructures of graphene. Nano resonators were made up in the mid-IR employing 15–80 nm width structures of graphene formed into the arrays. The experimental outcomes gave the capability to calculate the relations of dispersion of the graphene plasmons in the mid-IR spectral region. Additionally, the charge density of graphene was modified by applying the external voltage bias to the back-gated structures and this developed another mechanism for varying the resonances of plasmon (Currie et al., 2011). The shifting of the energy of plasmon as the function of applied bias and width are displayed in Figure 6.8.

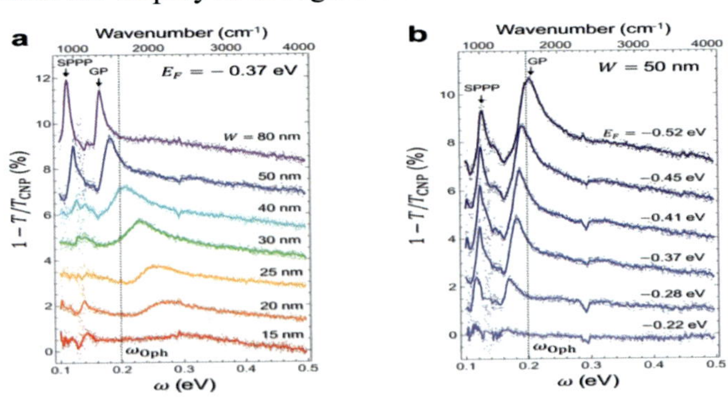

Figure 6.8. Optical transmission via plasmonic nano resonator of graphene arrays exhibits both GP (graphene plasmon) and SPPP (surface plasmon phonon

polariton) peaks. The vertical axis signifies the variation in the optical transmission standardized to optical transmission at the Dirac point, or CNP (charge neutral point). The resonances position change as the function of (a) width of a resonator; and (b) back-gate bias applied, displayed here as the shift in graphene's Fermi level, for the 50-nm wide structure.

One more plasmonic system of graphene was published, signifying an all-graphene photonic IC (integrated circuit). Generally, graphene is incorporated with semiconductors in order to develop photonic circuits, however, in this circumstance, only the graphene is employed. A substrate coated with polymer is used as the base in order to aid the graphene. The incident 1550-nanometer light is coupled to the graphene plasmonic waveguide and is directed to the graphene photodetector. The photonic system only displays a ~ 40-ms response time, conversely, this is the first all-graphene photonic IC (Xia et al., 2009; Mueller et al., 2010). This demonstration gives motivation for a novel class of cheap, 2D photonic ICs.

6.6. GRAPHENE AS A NONLINEAR OPTICAL DEVICE

Graphene has a close wavelength free linear absorption from the visible region via the far IR whereas absorbing the large amount (2.3%) of light for each monolayer. With the finite quantum states in each monolayer, graphene shows the nonlinear saturation of optical absorption at comparatively low saturation intensities. This is because of the principle of Pauli Exclusion, which allows only 2 Fermions to instantaneously occupy the single quantum state. Therefore, by Pauli's principle, probable absorption paths are congested, creating a saturation of absorption, thus making graphene more translucent under the high optical irradiance (Shiue et al., 2013).

Degenerate fs (femtosecond) optical pump-probe spectroscopy uses 2 optical pulses having the same wavelength with the large fluence pump pulse and the time-hindered lower fluence probe pulse in order to measure the temporal variation in the properties of a material. In order to measure the time behavior of saturable absorption in the graphene 800-nanometer-wavelength, 100-femtosecond optical pulses were normally applied in the experiment of degenerate optical pump-probe. Before the arrival of the pump pulse, there was not any variation in the transmission of probe pulse, however, for <100-femtosecond during the use of pump pulse, the sample of graphene upsurged its transmission by 2% when subjected to the 25-pJ pump pulse. Recovery of graphene to the constant-state transmission is defined by the 0.15-picosecond exponential decay. The measured intensities of saturation of 0.6–9 MW/cm^{2}are normal for the multilayer saturable absorbers of graphene (Figure 6.9) (Furchi et al., 2012).

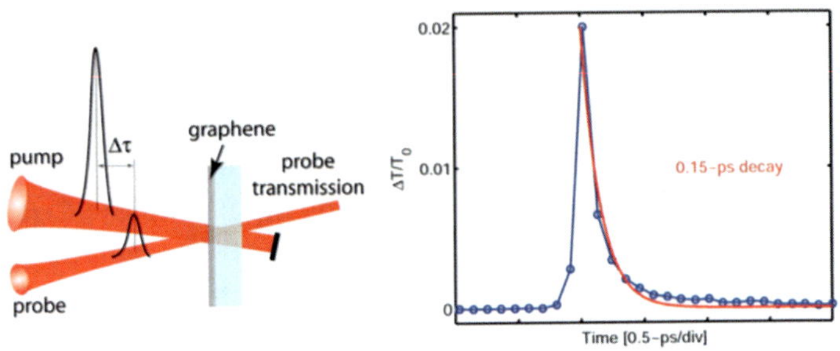

Figure 6.9. Optical pump-probe spectroscopy inspects the response of multilayer sample of graphene to the 100-femtosecond optical pulse and validates saturable absorption with the transient 2% upsurge in transmission, over the steady-state transmission.

Theory envisages the performance of graphene as the saturable absorber in the mid-IR and experiments illustrates the performance out to two-μm wavelengths. This is beneficial since comparatively few SESAMs (semiconductor saturable absorbers) have been made in the mid-IR region. The SESAMs utilized in this region are InGaAs, GaAs, InAs/GaSb, and GaInSb/GaSb (Xia et al., 2009; Mueller et al., 2010). Additionally, graphene has been stated to have larger depths of modulation and lower intensities of saturation than SESAMs (Xia et al., 2009; Mueller et al., 2010).

An additional useful characteristic of graphene for utilization as the nonlinear optical material is the high optical harm threshold (close to 100 GW/cm^2) of graphene to permit saturable absorption without the interest of harming the graphene. Additionally, graphene is a stretchy thin film and can compliantly coat materials. This permit bonding devices of graphene to optics to create incorporated nonlinear optical devices, like wavelength-independent SAM (saturable absorber mirrors). This is different to conventional semiconductor SAM, which are developed for particular wavelengths and are hard to develop at wavelengths longer than two μm. Therefore, graphene might play the future part as the SAM for making mode-locked lasers in the mid-infrared region (Shiue et al., 2013).

To examine the nonlinear susceptibility of graphene, THG (third-harmonic generation) experiments were carried out. Graphene displayed a strong THG when the excitation is normally in 3-photon resonance with graphene at M-point in the Brillouin zone. At the high energies, the band structure of graphene is not linear. The THG signal is over 100 times larger than in the glasses and is polarization discerning permitting the establishment of graphene's out-of-plane and in-plane nonlinear optical response. As the signal of THG is in the Ultraviolet spectral region, this gives an exclusive approach for identifying monolayers of graphene. One such easy graphene recognition system could utilize an 800-nanometer laser in order to stimulate the graphene and copy the area having an optical filter that congested 800-nanometer but passed the Ultraviolet wavelengths. The system would have practically no background scattered light as the excitation is in the near-infrared or visible region, and as there is comparatively little stray Ultraviolet light in the environment of the laboratory (Shiue et al., 2013).

REFERENCES

1. Adachi, S., (1985). GaAs, AlAs, and Al x Ga1- x As: Material parameters for use in research and device applications. *Journal of Applied Physics, 58*(3), R1–R29.
2. Avouris, P., & Freitag, M., (2013). Graphene photonics, plasmonics, and optoelectronics. *IEEE Journal of Selected Topics in Quantum Electronics, 20*(1), 72–83.
3. Bae, S., Kim, H., Lee, Y., Xu, X., Park, J. S., Zheng, Y., & Kim, Y. J., (2010). Roll-to-roll production of 30-inch graphene films for transparent electrodes. *Nature Nanotechnology, 5*(8), 574.
4. Bao, Q., & Loh, K. P., (2012). Graphene photonics, plasmonics, and broadband optoelectronic devices. *ACS Nano, 6*(5), 3677–3694.
5. Berger, C., Song, Z., Li, T., Li, X., Ogbazghi, A. Y., Feng, R., & De Heer, W. A., (2004). Ultrathin epitaxial graphite: 2D electron gas properties and a route toward graphene-based nanoelectronics. *The Journal of Physical Chemistry B, 108*(52), 19912–19916.
6. Bernardi, M., Palummo, M., & Grossman, J. C., (2013). Extraordinary sunlight absorption and one nanometer thick photovoltaics using two-dimensional monolayer materials. *Nano Letters, 13*(8), 3664–3670.
7. Bonse, J., Baudach, S., Krüger, J., Kautek, W., & Lenzner, M., (2002). Femtosecond laser ablation of silicon-modification thresholds and morphology. *Applied Physics A, 74*(1), 19–25.
8. Burke, P. J., Spielman, I. B., Eisenstein, J. P., Pfeiffer, L. N., & West, K. W., (2000). High frequency conductivity of the high-mobility two-dimensional electron gas. *Applied Physics Letters, 76*(6), 745–747.
9. Caldwell, J. D., Anderson, T. J., Culbertson, J. C., Jernigan, G. G., Hobart, K. D., Kub, F. J., & Myers-Ward, R. L., (2010). Technique for the dry transfer of epitaxial graphene onto arbitrary substrates. *ACS Nano, 4*(2), 1108–1114.
10. Chen, J. H., Jang, C., Xiao, S., Ishigami, M., & Fuhrer, M. S., (2008). Intrinsic and extrinsic performance limits of graphene devices on SiO_2. *Nature Nanotechnology, 3*(4), 206–209.
11. Chou, S. Y., Liu, Y., Khalil, W., Hsiang, T. Y., & Alexandrou, S., (1992). Ultrafast nanoscale metal-semiconductor-metal photodetectors on bulk and low-temperature grown GaAs. *Applied Physics Letters, 61*(7), 819–821.

12. Currie, M., Anderson, T. J., Wheeler, V. D., Nyakiti, L. O., Garces, N. Y., Myers-Ward, R. L., & Gaskill, D. K., (2013). Mode-locked 2-μm wavelength fiber laser using a graphene-saturable absorber. *Optical Engineering, 52*(7), 076101.

13. Currie, M., Caldwell, J. D., Bezares, F. J., Robinson, J., Anderson, T., Chun, H., & Tadjer, M., (2011). Quantifying pulsed laser induced damage to graphene. *Applied Physics Letters, 99*(21), 211909.

14. Currie, M., Quaranta, F., Cola, A., Gallo, E. M., & Nabet, B., (2011). Low-temperature grown GaAs heterojunction metal-semiconductor-metal photodetectors improve speed and efficiency. *Applied Physics Letters, 99*(20), 203502.

15. Eletskii, A. V., Iskandarova, I. M., Knizhnik, A. A., & Krasikov, D. N., (2011). Graphene: Fabrication methods and thermophysical properties. *Physics-Uspekhi, 54*(3), 227.

16. Furchi, M., Urich, A., Pospischil, A., Lilley, G., Unterrainer, K., Detz, H., & Mueller, T., (2012). Microcavity-integrated graphene photodetector. *Nano Letters, 12*(6), 2773–2777.

17. Geim, A. K., & Grigorieva, I. V., (2013). Van der Waals heterostructures. *Nature, 499*(7459), 419–425.

18. Grigorenko, A. N., Polini, M., & Novoselov, K. S., (2012). Graphene plasmonics. *Nature Photonics, 6*(11), 749–758.

19. Gupta, S., Frankel, M. Y., Valdmanis, J. A., Whitaker, J. F., Mourou, G. A., Smith, F. W., & Calawa, A. R., (1991). Sub picosecond carrier lifetime in GaAs grown by molecular beam epitaxy at low temperatures. *Applied Physics Letters, 59*(25), 3276–3278.

20. Ito, H., Kodama, S., Muramoto, Y., Furuta, T., Nagatsuma, T., & Ishibashi, T., (2004). High-speed and high-output InP-InGaAs unitraveling-carrier photodiodes. *IEEE Journal of Selected Topics in Quantum Electronics, 10*(4), 709–727.

21. Kamisan, A. I., Kamisan, A. S., Md Ali, R., Tunku, K. T. I., Hassan, O. H., Abdul, H. N., & Yahya, M. F., (2015). Synthesis of graphene via green reduction of graphene oxide with simple sugars. In: *Advanced Materials Research* (Vol. 1107, pp. 542–546). Trans Tech Publications Ltd.

22. Kim, J. T., Yu, Y. J., Choi, H., & Choi, C. G., (2014). Graphene-based plasmonic photodetector for photonic integrated circuits. *Optics Express, 22*(1), 803–808.

23. Kim, K. S., Zhao, Y., Jang, H., Lee, S. Y., Kim, J. M., Kim, K. S., & Hong, B. H., (2009). Large-scale pattern growth of graphene films for stretchable transparent electrodes. *Nature, 457*(7230), 706–710.
24. Koh, W. S., Gan, C. H., Phua, W. K., Akimov, Y. A., & Bai, P., (2013). The potential of graphene as an ITO replacement in organic solar cells: An optical perspective. *IEEE Journal of Selected Topics in Quantum Electronics, 20*(1), 36–42.
25. Lee, C., Wei, X., Kysar, J. W., & Hone, J., (2008). Measurement of the elastic properties and intrinsic strength of monolayer graphene. *Science, 321*(5887), 385–388.
26. Moon, J. S., & Gaskill, D. K., (2011). Graphene: Its fundamentals to future applications. *IEEE Transactions on Microwave Theory and Techniques, 59*(10), 2702–2708.
27. Mueller, T., Xia, F., & Avouris, P., (2010). Graphene photodetectors for high-speed optical communications. *Nature Photonics, 4*(5), 297–301.
28. Nair, R. R., Blake, P., Grigorenko, A. N., Novoselov, K. S., Booth, T. J., Stauber, T., & Geim, A. K., (2008). Fine structure constant defines visual transparency of graphene. *Science, 320*(5881), 1308–1308.
29. Nyakiti, D. G. Q. L., (2012). Formation of epitaxial graphene. *Graphene Nanoelectronics: From Materials to Circuits, 1*, 130–137.
30. Pan, W., Masuhara, N., Sullivan, N. S., Baldwin, K. W., West, K. W., Pfeiffer, L. N., & Tsui, D. C., (2011). Impact of disorder on the 5/2 fractional quantum hall state. *Physical Review Letters, 106*(20), 206806.
31. Shiue, R. J., Gan, X., Gao, Y., Li, L., Yao, X., Szep, A., & Englund, D., (2013). Enhanced photodetection in graphene-integrated photonic crystal cavity. *Applied Physics Letters, 103*(24), 241109.
32. Sun, D. M., Liu, C., Ren, W. C., & Cheng, H. M., (2013). A review of carbon nanotube-and graphene-based flexible thin-film transistors. *Small, 9*(8), 1188–1205.
33. VanMil, B. L., Myers-Ward, R. L., Tedesco, J. L., Eddy, C. R., Jernigan, G. G., Culbertson, J. C., & Gaskill, D. K., (2009). Graphene formation on SiC substrates. In: *Materials Science Forum* (Vol. 615, pp. 211–214). Trans Tech Publications Ltd.
34. Wallace, P. R., (1947). The band theory of graphite. *Physical Review, 71*(9), 622.

35. Wang, H. X., Wang, Q., Zhou, K. G., & Zhang, H. L., (2013). Graphene is light: Design, synthesis and applications of photo-active graphene and graphene-like materials. *Small, 9*(8), 1266–1283.
36. Wang, Q. H., Kalantar-Zadeh, K., Kis, A., Coleman, J. N., & Strano, M. S., (2012). Electronics and optoelectronics of two-dimensional transition metal dichalcogenides. *Nature Nanotechnology, 7*(11), 699–712.
37. Xia, F., Mueller, T., Lin, Y. M., Valdes-Garcia, A., & Avouris, P., (2009). Ultrafast graphene photodetector. *Nature Nanotechnology, 4*(12), 839–843.
38. Xing, G., Guo, H., Zhang, X., Sum, T. C., & Huan, C. H. A., (2010). The physics of ultrafast saturable absorption in graphene. *Optics Express, 18*(5), 4564–4573.
39. Yamashita, S., (2011). A tutorial on nonlinear photonic applications of carbon nanotube and graphene. *Journal of Lightwave Technology, 30*(4), 427–447.
40. Zhang, Y., Tan, Y. W., Stormer, H. L., & Kim, P., (2005). Experimental observation of the quantum hall effect and berry's phase in graphene. *Nature, 438*(7065), 201–204.

CHAPTER 7

Applications of Nanocomposites Based on Graphene in Supercapacitors

CONTENTS

7.1. INTRODUCTION

The research conducted on supercapacitor devices has concentrated heavily on the mechanical tractability of the solid-state appliances, with the objective of preserving high electrochemical performance whereas following the substantial drift of wearable and portable electronics becoming thin, small, flexible, and lightweight, which brings novel trials for the energy storage systems (Pushparaj et al., 2007; Arico et al., 2011). Several efforts to construct the lithe supercapacitor system based on graphene have been stated, without, and with the adding of binders.

At the glimpse, the making idea comes to an analogous point with the usage of a bendable supercapacitor electrode and the polymer gel electrolyte. Graphene now has become the crucial component in the creation of lithe electrodes due to its extraordinarily high mechanical strength, good conductivity, and outstanding surface area, which generally add extra benefits in averting the framework of an electrode from being broken as an outcome of the mechanical twisting/bending of the lithe device (Horng et al., 2010; Liu et al., 2010).

In contrast, the polymer gel electrolyte is frequently used as both an electrolyte reservoir and ion porous separator due to its semisolid characteristic of pores or microchannels all over the structure, which competently simplify the electrolyte ions flow whereas confirming no physical contact amongst the interface and electrodes (Endo et al., 2001).

Moreover, the frame of semisolid gives the maximum bond with negligible distance amongst the electrolyte/electrode/electrode interfaces, therefore efficiently improving the mechanism of charge storage. Usually, gel electrolytes based on PVA {poly(vinyl alcohol)} have been described in the literature, even though other polymers like potassium polyacrylate or the membrane separator (Celgard) available commercially have also been used. The technique of fabrication for the lithe solid-state supercapacitor gadget is briefed in Figure 7.1 (Lim et al., 2014).

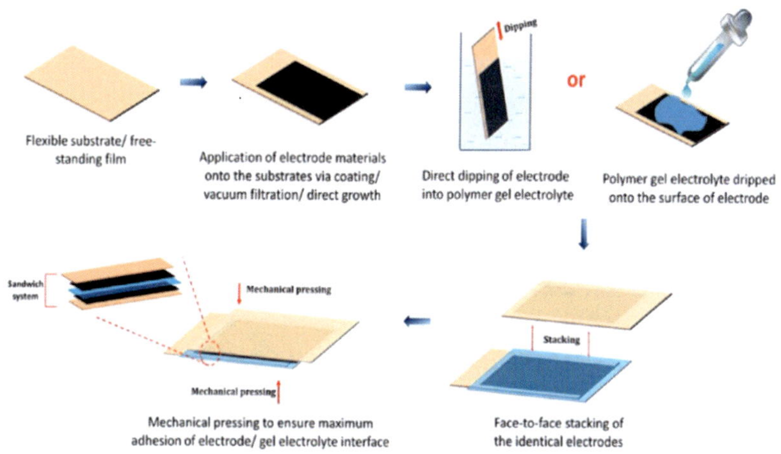

Figure 7.1. *Simplified diagram of creation of lithe supercapacitor device.*

Source: https://pubs.acs.org/doi/10.1021/acs.jpcc.5b10187.

7.2. SYNTHESIS METHOD FOR THE ELECTRODE MATERIALS

7.2.1. Electrodeposition/Electro-Polymerization

The technique of electrodeposition is among the most usual methods for the creation of supercapacitor materials for an electrode. This straightforward and facile synthesis method gives accurate governance of the thickness of consequential films along with the polymerization rate, which might be easily governed by the applied current density. Thus; this approach comprises only conditions of mild processing at room temperature without any lethal or surplus chemicals being used during the process (Stoller and Ruoff, 2010; Vidhyadharan et al., 2014).

This technique is generally used in the synthesis of usual conductive polymers, comprising PPy, PANI, polyacetylene, and PEDOT; conversely, the conductive polymers have similar restraint of mediocre mechanical strength, which outcomes in limited and low cycle stability (Zhang et al., 2014). Many alterations to the synthesis technique and ready materials have been stated to overpower this restraint comprising the instigation of materials based on EDLC like activated carbons, graphene, and CNTs, which normally have been stated to enhance the particular capacitance and whole cycle stability performance. For example, the creation of PANI nanowires (NWs) with the adding of SWCNT occasioned a substantial upsurge in the capacitive performance of the altered electrode (Zhou et al., 2014). Currently, the amendment trend concentrated on the use of graphene to improve the conductivity of polymers. Generally, graphene is the one-layer-thick planar sheet of carbon organized in the hexagonal manner, which provides many advantages like a large area, high capacity, and superb conductivity (Shi et al., 2014; Wang et al., 2014).

7.2.2. In Situ Polymerization

This technique frequently refers to a usual chemical polymerization method that has been broadly stated. Generally, the technique of polymerization comprises a reaction in the aqueous solution, through which the monomers normally disperse into the aqueous solution with the assistance of the process of sonication (Winter and Brodd, 2004). A particular oxidizing agent is usually added to start the process of polymerization inside the solution, and the sample is attained through the direct filtration method. Previously, the particular method was described to produce irregular granular masses with small ratios of nanofibers; conversely, with certain adjustments, it was currently stated that nanostructures like nanorods, nanofibers, and nanoparticles were attained with enhanced solution processability, as well as with better chemical and physical properties as compared to their bulk equivalents. Additionally, modifications, and enhancements of this conventional technique have also been stated elsewhere (Fan et al., 2011).

7.2.3. Direct Coating

This technique is among the most usual techniques used in the preparation of supercapacitor electrodes. Since the active materials are normally directly applied to the surface of the substrate itself, this technique relies heavily on the physical bond of electrode material on the substrate. Frequently, additive or binder like PVDF (polyvinylidene fluoride) and carbon black (CB) were used to give maximum adhesion, whereas preserving the electrical conductivity at the substrate interface/electrode material (Zhang et al., 2009; Geim and Novoselov, 2010). This was created in the form of slurry and physically directly coated onto the surface of the substrate.

7.2.4. Chemical Vapor Deposition (CVD)

This technique is applied normally when the permeability of the ultimate product is very critical. In the synthesis approach of graphene, chemical vapor deposition (CVD) is frequently stated as being capable to yield a defect-free structure of graphene that has a highly conductive network for the transfer of charge, with the lack of inter-sheet intersection resistance in the endless network; conversely, this technique normally needs a growth or template substrate to give a surface for the supplement of the freshly deposited layers of graphene (Novoselov et al., 2004, 2005). As stated by its name, the procedure is carried out under the vapor phase, through which the initiating material of the graphene is primarily made in vapor form, streamed, and subjected to high temperature (800°C to 1000°C), as well as the targeted substrate. The particular sample will thereby develop on the substrate with even and very fine morphology (Chee et al., 2016).

7.3. SUBSTRATE MATERIALS FOR LITHE SUPERCAPACITORS

Mechanical properties of the materials of the substrate play a critical role in the creation of a lithe supercapacitor electrode. Frequently, it provides mechanical strength and flexibility to the components of the electrode. Therefore, far, several materials of the substrate have been applied, comprising the most usually described, nickel foam, as well as the aluminum foil or foam, carbon cloths, and graphite sheets, primarily due to their outstandingly high conductivity, high porosity, and flexibility. The existence of the stable mechanical structure delivers a conductive background for the bond of the materials of an electrode, whereas also aiding the purpose as the current collector (Nair et al., 2008; Perera et al., 2013). Currently, the trend has concentrated on the use of CNFs (carbon nanofibers) in the preparation of supercapacitor electrodes. Indisputably, carbon nanofibers provide outstanding properties like high conductivity, mechanical flexibility, and very high particular surface area. Generally, carbon nanofibers have been synthesized through the process of polymer carbonization. Several kinds of polymers can be employed in order to form CNFs; conversely, PAN {poly(acrylonitrile)} and PVA {poly(vinyl alcohol)} are the most usual polymers employed in the synthesis, primarily due to their high content of carbon upon carbonization (Nyholm et al., 2011). Primarily, the polymer is usually dissolved in the solvent in order to form the polymer solution. The polymer is then imperiled to the process of electrospinning, during which the applied high potential (10 to 15 kV) ultimately forces the solution of polymer to form incessant fibers having an average diameter of various microns to nanometers, contingent on the applied potential. The polymerized nanofibers are introduced in the process of carbonization, where the matrix of polymer is imperiled to high temperatures in an inactive atmosphere. Ultimately, other organic remains will be ruined, leaving primarily carbon structures behind, thus creating endless carbon nanofibers. Numerous papers have also stated the utilization of carbon nanofiber in the creation of supercapacitor electrodes (Rogers et al., 2010; Jiang et al., 2012).

7.4. GRAPHENE NANOCOMPOSITE CENTERED ON KINDS OF ELECTRODE MATERIALS

7.4.1. Binder/Additives Added Electrodes

7.4.1.1. Graphene Electrodes/Additives

In the creation of the supercapacitor electrode, CB has frequently been used as the conductive additive, primarily due to its low cost and high conductivity (Stoller et al., 2008). As stated, a CB/graphene hybrid film was made and used directly as the supercapacitor electrode, for which rGO (reduced graphene oxide) was assorted with diverse amounts of CB and imperiled to vacuum filtration in order to attain the film of the nanocomposite. The CB/rGO film was free-standing and had flexible properties, as displayed in Figure 7.2(a) (Stoller et al., 2008; Huang et al., 2012). The existence of CB efficiently behaved as a spacer, which avoided the restacking of rGO layers; conversely, higher content of CB occasioned agglomeration (compare Figure 7.2(c) and (e)).

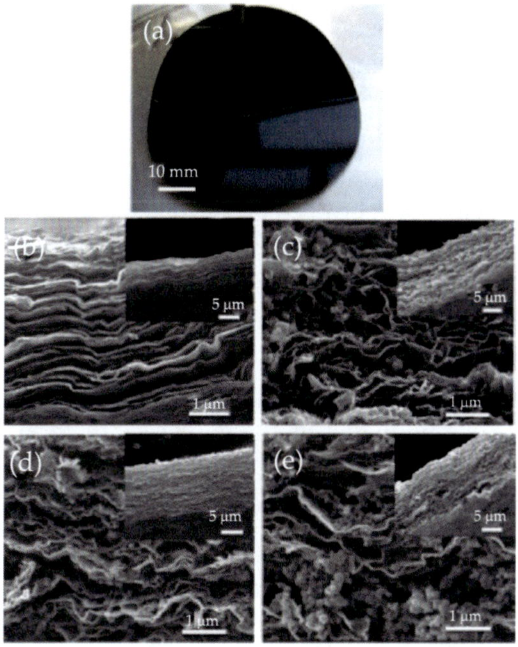

Figure 7.2. (a) Digital image of carbon black-2/rGO. SEM cross-sectional pictures of (b) rGO; (c) rGO/carbon black-1; (d) rGO/carbon black-1.5; and (e) rGO/carbon black-2, correspondingly.

Source: https://coek.info/pdf-graphene-carbon-black-hybrid-film-for-flexible-and-high-rate-performance-superca.html.

In terms of the electrochemical performance, the amalgam film was tested employing the 2-electrode configuration, through which two pieces of similar film and the film of cellulose ester filter were soaked in one M H_2SO_4 for two hours. The film of ester filter was then squeezed amongst the amalgam film and fastened by Ti (titanium) plates, which behaved as the current collectors (Wang et al., 2014). As anticipated, usual rectangular-shape CV (cyclic voltammetry) profiles were noted, which were accountable for the EDLC (electric double-layer capacitance). Stimulatingly, the rectangular CV shape persisted even at the high rate of a scan of 5 V/s, thus exhibiting the high-rate abilities. Likewise, a symmetrical triangular discharge/charge profile accredited to the EDLC was recorded. The made-up supercapacitor displayed extraordinary cycle stability by displaying 100% capacitance retention up to 2000 cycles. The enhancement was primarily accredited to the existence of CB, which efficiently behaved as spacers and avoided the restacking of layers of graphene and also enhanced the conductivity inside the nanocomposite film (Davies et al., 2011; Choi et al., 2012). Furthermore, the solid-state supercapacitor was made based on the hybrid rGO/CB-1.5 composite film. Two indistinguishable film pieces were rolled onto the Au-covered PET substrate, trailed by flaking off the membrane. A ready polymer gel electrolyte PVA/H_2SO_4 was dropped onto both of the electrodes; afterward, the 2 electrodes were pushed together for 10 hours (Gao et al., 2013). The polymer gel electrolyte behaved as the reservoir of electrolyte and the medium that permitted the electrodes in order to remain intact. Unusually, the CV curve of the made-up device displayed an outstanding CV rectangular shape without intervention in both the normal and bent states (Figure 7.3(a)), demonstrating the flawless structural reliability of the created device. A particular capacitance high as 79.6 F g^{-1} was noted at the fast rate of a scan of 1 V/s. The device's cycling stability was tested at the current density of around 2 A/g, with the first 3000 cycles verified in the normal state and the next 2000 cycles in the bent state. The capacitance preservation remained at 94%, even after the overall of 5000 cycles (Figure 7.3(b)). Its performance as the supercapacitor was confirmed further by fruitfully lighting the light-emitting diode (LED) circuit, as displayed in Figure 7.3(c) (Gao et al., 2012; Zhou et al., 2013).

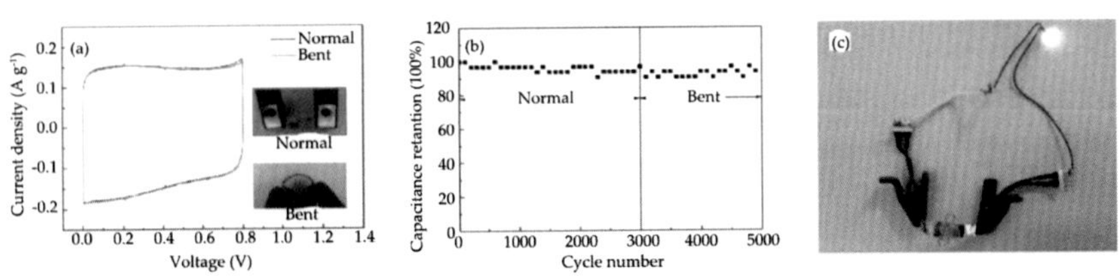

Figure 7.3. (a) Cyclic voltammetry curves of the solid-state supercapacitor on simple PET tested in normal and bent states; (b) cycle life of the solid-state supercapacitor on simple PET; (c) light-emitting diode lighted by 3 solid-state supercapacitors on simple PET in series.

Source: https://pubs.acs.org/doi/10.1021/acs.jpcc.5b10187.

7.4.1.2. Graphene Electrodes/Binder

In the preparation of the binder-added electrodes, mesoporous graphene was used by mixing 10 wt.% PTFE binder and 5 wt.% super-p, which was made into the coin-size cells of capacitor (Xie et al., 2014). The electrodes were disjointed through the Celgard porous membrane andEMIMBF_4 (1-ethyl-3-

methyldizolium tetrafluoroborate) as the organic electrolyte. The morphology of curved sheets of graphene specified a capability to avert close stacking. Thus, the mesoporous structure was preserved even after being compacted into the structure of the electrode. The size of the pore of the structure of graphene was inside the range of 2 to 25 nm (Xie et al., 2014). On the contrary, sheets of graphene made through traditional chemical reduction techniques have a higher inclination to restack with each other, which considerably decreased the effective surface area as the inter-graphene gaps were usually smaller than one nm, thus occasioning in a considerably smaller capacitance of almost <50 F g^{-1}. The importance of mesoporous structure controlled by the curved sheets of graphene was displayed by the high capacitance value of almost 100 to 250 F g^{-1} under the probable window of four V at the current density of one A/g (Wu et al., 2010).

7.4.2. Binder-Less Electrodes

Currently, the synthesis of these types of electrodes is described in various papers. The binder-less properties provide several advantages, comprising the cost reduction and reduction of electrical interference amongst the components at the interface (Yoo et al., 2011; Yang et al., 2014).

7.4.2.1. Pure Electrode of Graphene

It can be achieved by direct preparation of two-dimensional (2D) in-plane pristine graphene and RMGO (reduced multilayer graphene) in which the original graphene is synthesized through the technique of CVD and the multilayer films of graphene were synthesized through the chemical reduction of the graphene oxide (GO). The in-plane design comprised separating the large planar graphene sheet into 2 electrodes by forming the micrometer-sized gap physically through the scission on the layer of graphene and then popping the outer edges of the 2 electrodes with gold (Au) to permit them to behave as the collectors of current. An acidic electrolyte of polymer (PVA/H_3PO_4) was spread through the active surface and gap amongst the electrodes to act as the electrolyte reservoir. The amalgamation occasioned in the made-up device with flexible, optically transparent, and ultrathin features. The mass of the G electrode was very small to the fact that it inclines to cause incorrectness in the purpose of particular capacitance (Xie et al., 2013). Thus, the surface area of electrodes was considered, and the G electrode displayed the normalized capacitance of around 80 μF/cm centered on the curve of a CV; conversely, the capacitance of the RMGO device was almost 5 times higher, which was accredited to the manifold layered structure of decreased graphene that had adequate interlayer distance for optimum counter ions intercalation. None of the devices displayed performance deprivation, after 1500 discharge/charge cycles. An analogous supercapacitor was created with the traditional RMGO geometry employed for the material of the electrode. The stacked devices displayed a much lower particular capacitance of almost 140 μF^{-2}, which outlines that the promising component placement was crucial to exploit the efficient electrochemical area of the multilayer surface of graphene. In the meantime, a much easier approach comprising the laser reduction of oxide films of graphite to graphene has normally been revealed. The thin film of GO was primarily cast onto the stretchy substrate, which was trailed by direct laser engraving inside the Light Scribe DVD/CD optical drive available commercially (Chen et al., 2014).

A usual supercapacitor device was prepared by sandwiching the polymer gel electrolyte or an ion porous separator amongst 2 indistinguishable LSG electrodes (Figure 7.4(a)), which produced a thin device with a total thickness of nearly <100 μm. The device exhibited a perfect rectangular CV shape over the broad rate of a scan of 100 to 10,000 mV s^{-1}, with an almost triangular shape for CC curves attained at the high current density of nearly 10 A/g. In one M H_2SO_4 electrolyte, a supercapacitor displayed an area capacitance of nearly 4.04 mF/cm at one A/g and preserved an area capacitance of almost 1.84 mF/cm even at 1000 A/g. Moreover, the LSG supercapacitor engaged 96.5% of preliminary capacitance after 10,000 discharge/charge cycles. The created supercapacitor exhibited extraordinary elasticity, with the twisting of the device having no effect on the electrochemical performance, as represented by the CV curve exhibited in Figure 7.4(b) (El-Kady et al., 2012).

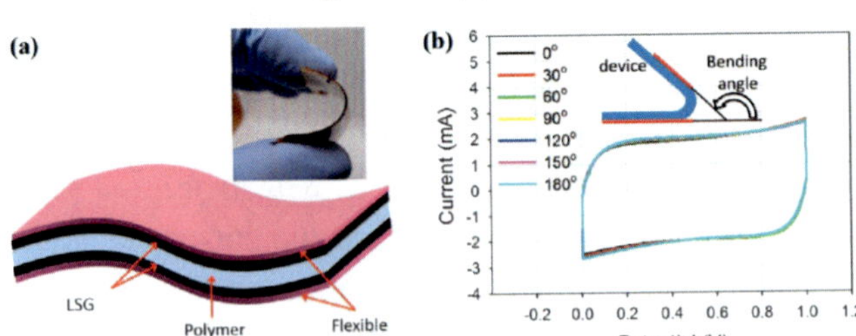

Figure 7.4. (a) Design and preparation of stretchy, all-solid-state light scribe graphene electrochemical capacitor; (b) winding the device did not affect the performance, as observed in the CVs gathered at the scan rate of around 1000 mV/s(He et al., 2013).

Source: https://science.sciencemag.org/content/335/6074/1326.

Rather than using the pure solid film, instead, a unique method was used to create the graphene hydrogel for usage as the flexible electrode in the solid-state supercapacitor (He et al., 2013).

The hydrogel of graphene was made through the altered hydrothermal reduction approach and had a thickness of nearly three mm. It was engrossed in the one M H_2SO_4 aqueous solution. The hydrogel of graphene was particularly cut and pressed on the gold-covered polyimide substrate with the pressure of nearly one MPa to accomplish the thin film having a real mass of almost 2 mg/cm^2.

Comparable to former reports, the solid-state supercapacitor was prepared by sandwiching the PVA/H_2SO_4 polymer electrolyte layer amongst indistinguishable electrodes created of graphene hydrogel/film of polyimide, with an extra mechanical pressure of one MPa for nearly 30 min, occasioning in a thin separator layer being developed amongst the electrodes (He et al., 2013). The simplified diagram and pictures of the prepared supercapacitor device are exhibited in Figure 7.5(a) and (b), correspondingly.

The morphology of three-dimensional (3D) graphene framework was maintained well, with an endless porous network visible still (Figure 7.5(c)–(f)), even after the mechanical pressing, which occasioned in the reduction in the general thickness of film from nearly 3 mm–120 μm; though a large dissimilarity in electrical conductivities was noted before and after pressing.

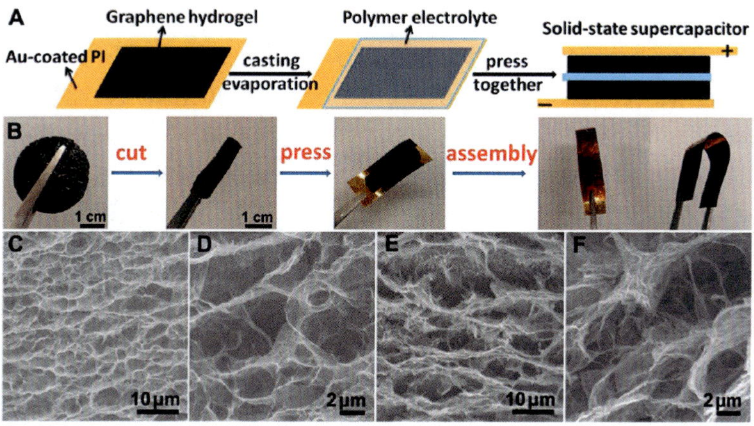

Figure 7.5. (a) Simplified diagram; and (b) images of the process of fabrication for stretchy solid-state supercapacitors centered on the graphene hydrogel films; (c) low; and (d) high magnification scanning electron microscope pic-

tures of the inner microstructure of graphene hydrogel before compressing; (e) low; and (f) high magnification scanning electron microscope pictures of the inner microstructure of graphene hydrogel film after compressing.

Source: https://www.sciencedirect.com/science/article/abs/pii/S0008622313002947.

Just like the former reports, a CV loop near to the perfect rectangular shape was documented for the as-created supercapacitor, at the high rate of a scan of 200 mV/s, which normally reflected an outstanding capacitive behavior with the low contact resistance. The uppermost particular capacitance was measured to be 187 F/g at the scan rate of almost 10 mV/s. A nearly even triangular shape for the galvanostatic discharge/charges curve was noted, for which the particular capacitance was measured to be nearly 186 F/g at the current density of one A/g and dropped to almost 130 F/g at the fast current of 20 A/g. The conclusions further established the outstanding conductive network of the graphene hydrogel with quite a large particular surface area, whereas penetrating the electrolyte into the network of the electrode.

7.4.2.2. Symmetrical Supercapacitor

Graphene/PANI (polyaniline) having a wavy shape was created as the flexible supercapacitor electrode through the series of chemical growing approaches, during which the Ni (nickel) foam template was entirely eliminated, leaving behind only the spongy graphene template and the sheet of graphene was consequently pulse electrodeposited with polyaniline. The flexible supercapacitor device was created by dipping the indistinguishable electrodes in the PVA/H_3PO_4 polymer electrolyte, trailed by assembling the electrodes. On finishing the solidification of polymer gel, the device was entrenched into the aqueous solution of Ecoflex and then cured. The spongy graphene had a crumpled structure, with some of the cracks recognized, as exhibited in Figure 7.6(a) and (b).

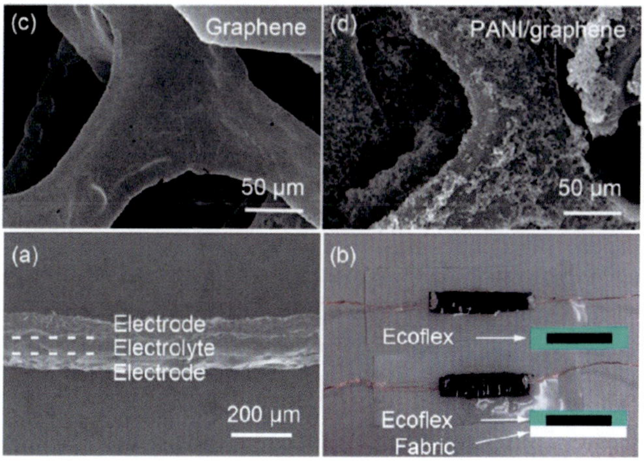

Figure 7.6. (a) Cross-sectional scanning electron microscope picture of the device; (b) flexible supercapacitors condensed in Ecoflex/fabric and Eco flex. Scanning electron microscope pictures of (c) porous graphene; and (d) Polyaniline I/graphene.

Source: https://pubs.rsc.org/en/content/articlelanding/2014/ta/c4ta00734d.

REFERENCES

1. Arico, A. S., Bruce, P., Scrosati, B., Tarascon, J. M., & Van, S. W., (2011). Nanostructured materials for advanced energy conversion and storage devices. In: *Materials for Sustainable Energy: A Collection of Peer-Reviewed Research and Review Articles from Nature Publishing Group* (Vol. 1, pp. 148–159).
2. Chee, W. K., Lim, H. N., Zainal, Z., Huang, N. M., Harrison, I., & Andou, Y., (2016). Flexible graphene-based supercapacitors: A review. *The Journal of Physical Chemistry C, 120*(8), 4153–4172.
3. Chen, Q., Meng, Y., Hu, C., Zhao, Y., Shao, H., Chen, N., & Qu, L., (2014). MnO_2-modified hierarchical graphene fiber electrochemical supercapacitor. *Journal of Power Sources, 247*, 32–39.
4. Choi, B. G., Chang, S. J., Kang, H. W., Park, C. P., Kim, H. J., Hong, W. H., & Huh, Y. S., (2012). High performance of a solid-state flexible asymmetric supercapacitor based on graphene films. *Nanoscale, 4*(16), 4983–4988.
5. Davies, A., Audette, P., Farrow, B., Hassan, F., Chen, Z., Choi, J. Y., & Yu, A., (2011). Graphene-based flexible supercapacitors: Pulse-electropolymerization of polypyrrole on free-standing graphene films. *The Journal of Physical Chemistry C, 115*(35), 17612–17620.
6. El-Kady, M. F., Strong, V., Dubin, S., & Kaner, R. B., (2012). Laser scribing of high-performance and flexible graphene-based electrochemical capacitors. *Science, 335*(6074), 1326–1330.
7. Endo, M., Takeda, T., Kim, Y. J., Koshiba, K., & Ishii, K., (2001). High power electric double-layer capacitor (EDLC's); from operating principle to pore size control in advanced activated carbons. *Carbon Letters, 1*(3, 4), 117–128.
8. Fan, Z., Yan, J., Wei, T., Zhi, L., Ning, G., Li, T., & Wei, F., (2011). Asymmetric supercapacitors based on graphene/MnO_2 and activated carbon nanofiber electrodes with high power and energy density. *Advanced Functional Materials, 21*(12), 2366–2375.
9. Gao, H., Xiao, F., Ching, C. B., & Duan, H., (2012). Flexible all-solid-state asymmetric supercapacitors based on free-standing carbon nanotube/graphene and Mn_3O_4 nanoparticle/graphene paper electrodes. *ACS Applied Materials &and Interfaces, 4*(12), 7020–7026.
10. Gao, K., Shao, Z., Li, J., Wang, X., Peng, X., Wang, W., & Wang,

F., (2013). Cellulose nanofiber-graphene all-solid-state flexible supercapacitors. *Journal of Materials Chemistry A, 1*(1), 63–67.

11. Geim, A. K., & Novoselov, K. S., (2010). The rise of graphene. In: *Nanoscience and Technology: A Collection of Reviews from Nature Journals* (Vol. 1, pp. 11–19).
12. He, Y., Chen, W., Li, X., Zhang, Z., Fu, J., Zhao, C., & Xie, E., (2013). Freestanding three-dimensional graphene/MnO_2 composite networks as ultralight and flexible supercapacitor electrodes. *ACS Nano, 7*(1), 174–182.
13. Horng, Y. Y., Lu, Y. C., Hsu, Y. K., Chen, C. C., Chen, L. C., & Chen, K. H., (2010). Flexible supercapacitor based on polyaniline nanowires/carbon cloth with both high gravimetric and area-normalized capacitance. *Journal of Power Sources, 195*(13), 4418–4422.
14. Huang, Y., Liang, J., & Chen, Y., (2012). An overview of the applications of graphene-based materials in supercapacitors. *Small, 8*(12), 1805–1834.
15. Jiang, J., Li, Y., Liu, J., Huang, X., Yuan, C., & Lou, X. W., (2012). Recent advances in metal oxide-based electrode architecture design for electrochemical energy storage. *Advanced Materials, 24*(38), 5166–5180.
16. Lim, Y. S., Tan, Y. P., Lim, H. N., Huang, N. M., Tan, W. T., Yarmo, M. A., & Yin, C. Y., (2014). Potentiostatically deposited polypyrrole/graphene decorated nano-manganese oxide ternary film for supercapacitors. *Ceramics International, 40*(3), 3855–3864.
17. Liu, C., Yu, Z., Neff, D., Zhamu, A., & Jang, B. Z., (2010). Graphene-based supercapacitor with an ultrahigh energy density. *Nano Letters, 10*(12), 4863–4868.
18. Nair, R. R., Blake, P., Grigorenko, A. N., Novoselov, K. S., Booth, T. J., Stauber, T., & Geim, A. K., (2008). Fine structure constant defines visual transparency of graphene. *Science, 320*(5881), 1308–1308.
19. Novoselov, K. S., Geim, A. K., Morozov, S. V., Jiang, D., Zhang, Y., Dubonos, S. V., & Firsov, A. A., (2004). Electric field effect in atomically thin carbon films. *Science, 306*(5696), 666–669.
20. Novoselov, K. S., Jiang, D., Schedin, F., Booth, T. J., Khotkevich, V. V., Morozov, S. V., & Geim, A. K., (2005). Two-dimensional atomic crystals. *Proceedings of the National Academy of Sciences, 102*(30), 10451–10453.

21. Nyholm, L., Nyström, G., Mihranyan, A., & Strømme, M., (2011). Toward flexible polymer and paper-based energy storage devices. *Advanced Materials, 23*(33), 3751–3769.
22. Perera, S. D., Rudolph, M., Mariano, R. G., Nijem, N., Ferraris, J. P., Chabal, Y. J., & Balkus, Jr. K. J., (2013). Manganese oxide nanorod-graphene/vanadium oxide nanowire-graphene binder-free paper electrodes for metal oxide hybrid supercapacitors. *Nano Energy,* 2(5), 966–975.
23. Pushparaj, V. L., Shaijumon, M. M., Kumar, A., Murugesan, S., Ci, L., Vajtai, R., & Ajayan, P. M., (2007). Flexible energy storage devices based on nanocomposite paper. *Proceedings of the National Academy of Sciences, 104*(34), 13574–13577.
24. Rogers, J. A., Someya, T., & Huang, Y., (2010). Materials and mechanics for stretchable electronics. *Science, 327*(5973), 1603–1607.
25. Shi, Y., Pan, L., Liu, B., Wang, Y., Cui, Y., Bao, Z., & Yu, G., (2014). Nanostructured conductive polypyrrole hydrogels as high-performance, flexible supercapacitor electrodes. *Journal of Materials Chemistry A, 2*(17), 6086–6091.
26. Stoller, M. D., & Ruoff, R. S., (2010). Best practice methods for determining an electrode material's performance for ultracapacitors. *Energy and Environmental Science, 3*(9), 1294–1301.
27. Stoller, M. D., Park, S., Zhu, Y., An, J., & Ruoff, R. S., (2008). Graphene-based ultracapacitors. *Nano Letters, 8*(10), 3498–3502.
28. Vidhyadharan, B., Zain, N. K. M., Misnon, I. I., Abd, A. R., Ismail, J., Yusoff, M. M., & Jose, R., (2014). High performance supercapacitor electrodes from electrospun nickel oxide nanowires. *Journal of Alloys and Compounds, 610,* 143–150.
29. Wang, Y. S., Li, S. M., Hsiao, S. T., Liao, W. H., Chen, P. H., Yang, S. Y., & Hu, C. C., (2014). Integration of tailored reduced graphene oxide nanosheets and electrospun polyamide-66 nanofabrics for a flexible supercapacitor with high-volume-and high-area-specific capacitance. *Carbon, 73*, 87–98.
30. Wang, Y., Chen, J., Cao, J., Liu, Y., Zhou, Y., Ouyang, J. H., & Jia, D., (2014). Graphene/carbon black hybrid film for flexible and high-rate performance supercapacitor. *Journal of Power Sources, 271*, 269–277.

31. Winter, M., & Brodd, R. J., (2004). *What are Batteries, Fuel Cells, and Supercapacitors?* (Vol. 1, pp. 1–22).
32. Wu, Q., Xu, Y., Yao, Z., Liu, A., & Shi, G., (2010). Supercapacitors based on flexible graphene/polyaniline nanofiber composite films. *ACS Nano, 4*(4), 1963–1970.
33. Xie, J., Sun, X., Zhang, N., Xu, K., Zhou, M., & Xie, Y., (2013). Layer-by-layer β-Ni$(OH)_2$/graphene nanohybrids for ultra-flexible all-solid-state thin-film supercapacitors with high electrochemical performance. *Nano Energy, 2*(1), 65–74.
34. Xie, Y., Liu, Y., Zhao, Y., Tsang, Y. H., Lau, S. P., Huang, H., & Chai, Y., (2014). Stretchable all-solid-state supercapacitor with wavy shaped polyaniline/graphene electrode. *Journal of Materials Chemistry A, 2*(24), 9142–9149.
35. Yang, C., Shen, J., Wang, C., Fei, H., Bao, H., & Wang, G., (2014). All-solid-state asymmetric supercapacitor based on reduced graphene oxide/carbon nanotube and carbon fiber paper/polypyrrole electrodes. *Journal of Materials Chemistry A, 2*(5), 1458–1464.
36. Yoo, J. J., Balakrishnan, K., Huang, J., Meunier, V., Sumpter, B. G., Srivastava, A., & Ajayan, P. M., (2011). Ultrathin planar graphene supercapacitors. *Nano Letters, 11*(4), 1423–1427.
37. Zhang, D., Yan, H., Lu, Y., Qiu, K., Wang, C., Tang, C., & Luo, Y., (2014). Hierarchical mesoporous nickel cobaltite nanoneedle/carbon cloth arrays as superior flexible electrodes for supercapacitors. *Nanoscale Research Letters, 9*(1), 1–9.
38. Zhang, Y., Sun, X., Pan, L., Li, H., Sun, Z., Sun, C., & Tay, B. K., (2009). Carbon nanotube-zinc oxide electrode and gel polymer electrolyte for electrochemical supercapacitors. *Journal of Alloys and Compounds, 480*(2), L17–L19.
39. Zhou, S., Zhang, H., Zhao, Q., Wang, X., Li, J., & Wang, F., (2013). Graphene-wrapped polyaniline nanofibers as electrode materials for organic supercapacitors. *Carbon, 52*, 440–450.
40. Zhou, Z., Wu, X. F., & Hou, H., (2014). Electrospun carbon nanofibers surface-grown with carbon nanotubes and polyaniline for use as high-performance electrode materials of supercapacitors. *RSC Advances, 4*(45), 23622–23629.

CHAPTER 8

Fundamentals of Graphene Transistors

CONTENTS

8.1. INTRODUCTION

From time to time, a distinctive article initiates a revolution in science and technology. In October 2004, such revolution was initiated, when condensed matter physicists stated that they had made graphene two-dimensional (2D) sheets of carbon atoms and perceived the electric field impact in their samples (Novoselov et al., 2004). It was not extended earlier this novel material attracted the consideration of the electron-device community, and currently, an increasing number of groups are effectively fabricating graphene transistors. Key chip-makers are currently active in graphene research and the International Technology Roadmap for the Semiconductors, the tactical planning document for the semiconductor industry, take the graphene to be amongst the candidate materials for the post-silicon electronics (Grau, 2011).

Numerous excellent analyzes on the basic science of graphene had been published in the latest years. Given the enhancing attention in graphene in the electron device community and continuing discussions of the prospective of graphene transistors, an analysis article concentrating on graphene devices is timely (Geim and Novoselov, 2010). Now, from the point of view of a device engineer, I debate the potential of graphene as a novel material for electron devices and review the state of the art for graphene transistors. I would focus mainly on the FET (field-effect transistor), as this is the most effective device concept in electronics and since most effort on graphene devices till yet had been associated with FETs (Geim, 2009; Neto et al., 2009).

Two basic divisions of semiconductor electronics are radiofrequency devices and digital logic devices. The extent of readiness to introduce novel device concepts is usually higher for radiofrequency applications, in part due to the fortunes of digital logic rely almost completely on the performance of a single kind of device: the MOSFET (silicon metal-oxide-semiconductor FET). For decades, producing MOSFETs smaller had been key to the development of digital logic. This size scaling had allowed the complexity of integrated circuits (ICs) to become double after every 18 months, leading to substantial improvements in performance and reduces in price per transistor (Chau et al., 2007; Schwierz et al., 2010). Currently, processors comprising 2 billion MOSFETs, numerous with gate lengths of just 30 nm, are in bulk production (Figure 8.1).

As the fabrication of ICs is extremely complicated, semiconductor fabrication plants are highly expensive (at current costing several billion US dollars). Moreover, because scaling alone had given the needed performance developments from one generation to the next of ICs, there had been a slight motivation for the chip manufacturers to introduce devices founded on a material other than silicon or a fundamentally diverse physics (Schwierz and Liou, 2003).

Though there is a consensus in the society that MOSFET scaling is forthcoming its limits and that, in the long run, it would be essential to introduce novel material and device concepts to confirm that performance continues to increase (Ning and Taur, 1998).

The condition is different for radiofrequency electronics. This field was led by defense applications till the late 1980s, and though it progressed into the mainstream in the 1990s due to developments in wireless communications, the military constantly provides substantial financial assistance for research into novel radio-frequency devices. This, collected with the fact that radiofrequency circuits are far less complicated than digital logic chips, had led to manufacturers of radiofrequency chips being more vulnerable to novel device concepts. A sign of this is the huge variety of different transistor kinds and materials utilized in radiofrequency electronics: these comprise HEMTs (high-electron-mobility transistors) founded on iii–v semiconductors like InP and GaAs, MOSFETs, silicon n-channel, and different kinds of bipolar transistor (Frank et al., 1998).

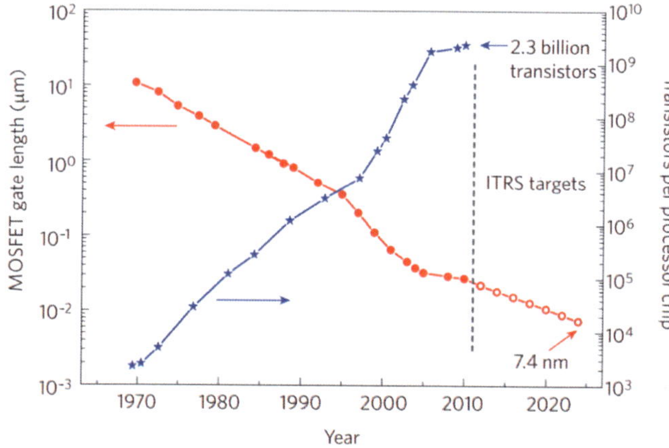

Figure 8.1. Trends in digital electronics.

Source: https://ieeexplore.ieee.org/document/6518205/.

As it is discussed below, graphene is possibly well suitable for radiofrequency applications due to its favorable carrier transport properties and its completely 2D structure. This, joined with the relative openness of the radiofrequency-electronics manufacturing to novel materials, proposes that graphene might be made its first presence in radiofrequency applications rather than in the logic circuits (Hughes and Tasker, 1989; Uyemura, 1999).

8.2. FET PHYSICS

A FET comprises a gate, a channel region joining drain and source electrodes, and a barrier splitting the gate from the channel (Figure 8.2(a)). The process of a conventional FET depends on regulating the channel conductivity, and therefore the drain current, by a voltage, V_{GS}, applied amongst the source and gate (Nguyen et al., 1989).

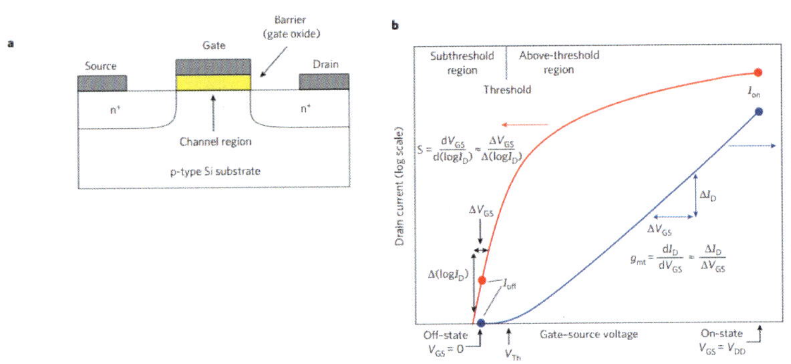

Figure 8.2. Conventional FETs. (a) cross-section of an n-channel Si MOSFET; (b) FET relocates characteristics display ID (on a linear scale on the right and a logarithmic scale on the left) vs. the VGS, gate-source voltage.

Source: https://ieeexplore.ieee.org/document/6518205/.

Note: For high-speed applications, FETs must respond rapidly to the variations in V_{GS}; this needs fast carriers and short gates in the channel. Inappropriately, FETs with short gates regularly suffer from degraded electrostatics and additional problems (mutually known as short channel impacts), like as drain made barrier lowering, threshold-voltage roll-off, and reduced drain-current saturation (Thompson et al., 2005). Scaling theory forecasts that a FET with a thin gate-controlled region and a thin barrier (measured in the vertical direction in Figure 8.2(a)) would be robust beside short-channel impacts down to very small gate lengths (measured in the horizontal direction in Figure 8.2(a)). The probability of having channels that are only one atomic layer thick is possibly the most striking feature of graphene for utilization in transistors. (Mobility, which is frequently considered to be graphene's most beneficial property for applications in nanoelectronics, is deliberated later.) By contrast, the channels in iii–v HEMTs are usually 10–15 nm thick, and though silicon on insulator MOSFETs with channel (that is, silicon body) thicknesses of fewer than 2 nm had been stated, rough interfaces affected their mobility to worsen. More significantly, the body of these MOSFETs displayed thickness variations that would lead to unacceptably huge threshold-voltage variations (and the same problems are expected to happen when the thickness of the channel in a iii–v HEMT is decreased to only a few nanometers). These issues happen at thicknesses that are various times greater than the thickness of graphene (Aberg and Hoyt, 2005).

The series resistances amongst the source and the channel and drain terminals are also significant, and their adverse effect on the FET becomes more distinct as the gate length reduces. Therefore, device engineers devote significant effort to developing transistor designs in which short-channel impacts are suppressed and series resistances are decreased (Boehm et al., 1962; May, 1969).

Recent digital logic is founded on silicon CMOS (complementary metal-oxide-semiconductor) technology. CMOS logic gates comprise of both p- and n-channel MOSFETs that could switch amongst the on-state (with a huge on-current, I_{on}, and $V_{GS} = \pm V_{DD}$, where V_{DD} is the extreme voltage delivered to the device) and the off-state (with a little off-current, I_{off}, and $V_{GS} = 0$). In the digital logic terminology, a gate is not the gate terminal of a transistor however a blend of two or more transistors that could perform a specific logic operation. The value of V_{GS} at which the FET is just on the edge of switching on is the threshold voltage, V_{th}. Figure 8.2(b) demonstrates the transfer features of an n-channel FET specifying the off-state and the on-state. Beneficial measures with which to evaluate the switching behavior are the subthreshold swing, S (related to the sub-threshold region), and the terminal transconductance, g_{mt} (related to the above-threshold region).

In the stable state, a definite number of the MOSFETs in a CMOS logic gate are constantly switched off so that no current excluding the small I_{off} pass through the gate (Van Bommel et al., 1975). The capability of silicon MOSFETs to switch off permits silicon CMOS to give tremendously low static power dissipation (which is the cause why silicon CMOS had overcome all competing logic technologies). Therefore, any successor to the silicon MOSFET that is to be used in CMOS-like logic must have outstanding switching competencies, as well as an on-off ratio, I_{on}/I_{off}, of between 10^4 and 10^7. In a conventional FET, this needs semiconducting channels with an ample bandgap, preferably 0.4 eV or more. Furthermore, p-, and n-channel FETs with balanced threshold voltages, that is, with $V_{Th,n} = -V_{Th,p}$, are required for proper CMOS operation (Figure 8.3) (Van Bommel et al., 1975).

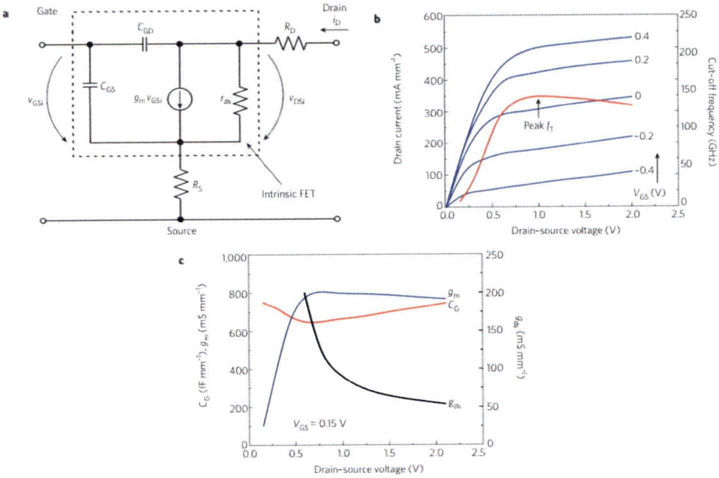

Figure 8.3. FET (direct current) and small-signal operation. (a) Small-signal same like FET circuit; (b) the ID (blue lines), drain current, at diverse values of VGS, and the fT (red line), cut-off frequency, both vs. VDS for a radiofrequency Ga As high electron-mobility transistor15,16; (c) the intrinsic transconductance (blue line), the whole gate capacitance, CG = CGS + CGD (red line), and the gds (1/rds; black line), and the drain conductance, vs. VDS for the similar FET.

Source: https://ieeexplore.ieee.org/document/6518205/.

8.3. GRAPHENE PROPERTIES RELEVANT TO TRANSISTORS

Single-layer graphene is a completely 2D material. Its lattice comprises of regular hexagons with a carbon atom at every corner. The bond length amongst the lattice constant, *a*, is 2.46 Å and the adjacent carbon atoms, L_b, is 1.42 Å (Figure 8.4(a)). The initial reports on this material seemed decades ago, even earlier to the name graphene had been devised, however, it took the pioneering 2004 paper through the Manchester group to spark the current explosion of interest in the material (Kim et al., 2009).

Currently, the most widespread approaches to graphene making are growth on metals, mechanical exfoliation (Kim et al., 2009), and subsequent graphene transform to insulating substrates, and thermal decomposition of SiC to form so-termed epitaxial graphene on the upper of SiC wafers (Berger et al., 2006; Reina et al., 2009). Exfoliation is yet popular for laboratory utilization; however it is not suitable to the electronics industry, while the other two choices both had the potential for generating wafer-scale graphene. After the graphene had been formed, common semiconductor processing methods (like metallization, lithography, and etching) could be applied to fabricate graphene transistors (Han et al., 2007; Kedzierski et al., 2008).

8.3.1. Bandgap

Huge-area graphene is a semimetal with zero bandgaps. Its conduction bands (CBs) and valence are cone-shaped and come across at the K points of the Brillouin zone (Figure 8.4(b)). As the bandgap is zero, devices with channels formed of large-area graphene cannot be switched off and thus are not appropriate for logic applications. Though the band structure of graphene could be modified, and it is probable to open a bandgap in 3 ways: by restraining huge-area graphene in one dimension to create graphene nanoribbons (GNRs), through biasing bilayer graphene, and through applying strain to graphene (Li et al., 2008; Young et al., 2009).

Both edges, zigzag nanoribbons, and armchair nanoribbons (the two perfect kinds of nanoribbon; Figure 8.4(a)) had a bandgap that is, to a good estimation, inversely proportional to the thickness of the nanoribbon. The opening of a bandgap in nanoribbons had been confirmed experimentally for widths down to around 1 nm, and experiments and theory both disclose bandgaps in an additional of 200 meV for widths below 20 nm (Figure 8.4(c)). Still, it should be noted that real nanoribbons had rough widths and edges that alter along their lengths. Even uncertain edge disorder eradicates any difference in the bandgap amongst nanoribbons with altered edge geometries, doping, and edge functionalization could also affect the bandgap (Chen et al., 2007; Yang et al., 2007).

8.3.2. Mobility

The most commonly stated benefit of graphene is its high carrier mobility at room temperature. Mobilities of 10 k–15 k $cm^2 V^{-1} s^{-1}$ is regularly measured for the exfoliated graphene on SiO_2-covered silicon wafers, and higher limits of amongst 40 k and 70 k $cm^2 V^{-1} s^{-1}$. Furthermore, in the absence of ripples and charged impurities, mobilities of 200 k $cm^2 V^{-1} s^{-1}$ had been predicted, and mobility of 10^6 $cm^2 V^{-1} s^{-1}$. For huge area graphene grownup on nickel and moved to a substrate, mobilities larger than 3,700 $cm^2 V^{-1} s^{-1}$ had been measured (Castro et al., 2007; Evaldsson et al., 2008).

Lastly, for epitaxial graphene on silicon carbide (SiC), the mobility relies on whether the graphene is grownup on the carbon face or the silicon face of SiC. Though graphene grownup on the carbon face had greater mobility (values of ~5,000 $cm^2 V^{-1} s^{-1}$ related with ~1,000 $cm^2 V^{-1} s^{-1}$ for graphene grownup on the silicon face), it is easier to grow bilayer and single-layer graphene on the silicon face, which creates the silicon face of SiC more suitable for electronic applications (Ohta et al., 2006; Gava et al., 2009).

In initial graphene MOS systems, the mobility was affected through the utilization of a top-gate dielectric (Zhang et al., 2009). Though the latest demonstration of mobilities of about 23 k $cm^2 V^{-1} s^{-1}$ in top-gated graphene MOS channels and the observation of same mobilities earlier and after top-gate formation display that high-mobility graphene MOS channels could be formed with an appropriate choice of the optimization and gate dielectric of the deposition procedure (Rotenberg et al., 2008).

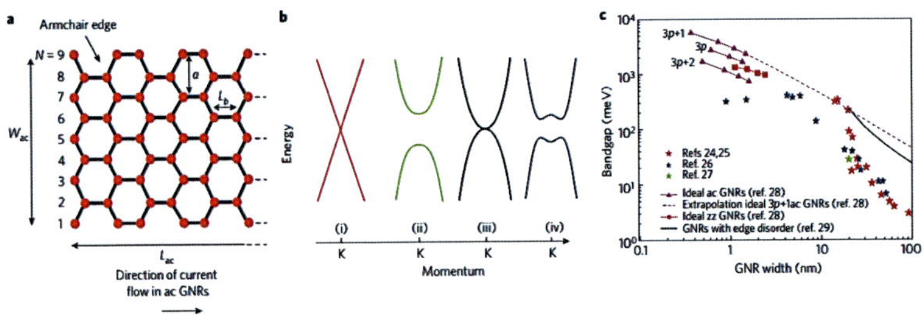

Figure 8.4. Properties of graphene nanoribbons and graphene. (a) Schematic of an ac(armchair) (ac) graphene nanoribbon (GNR) of width Wac and length Lac; (b) band structure about the K point of (i) unbiased bilayer graphene; (ii) bilayer graphene with an applied perpendicular field; (iii) large-area graphene; and (iv) graphene nanoribbons; (c) bandgap vs. nanoribbon width from calculations and experiments.

Source: https://journals.aps.org/prb/abstract/10.1103/PhysRevB.78.161407.

These mobility numbers are striking; however, they need closer inspection. The high mobilities stated above relate to huge-area graphene, which is gapless. A common trend for the conventional semiconductors is that the electron mobility reduces as the bandgap rises, and the same trend had been predicted for GNRs and carbon nanotubes (CNTs) (Figure 8.5(a)). This means that mobility in the nanoribbons with a bandgap equivalent to that of silicon (1.1 eV) is predicted to be lower than in bulk silicon and not greater than the mobility in the silicon channel of a conventional MOS device (Zhou et al., 2007; Kim et al., 2008).

8.3.3. High-Field Transport

In the days when FETs had gated many micrometers long, mobility was a suitable measure of the swiftness of carrier transport. Exactingly speaking, though, the mobility describes carrier transport in small electric fields; the little gate lengths in modern FETs consequences in high fields in a sizeable part of the channel, decreasing the significance of mobility to device performance (Chen et al., 2007; Peng and Ahuja, 2008). To demonstrate this, let us take a drain-source voltage of 1 V FET and a gate 100 nm long. If we suppose a voltage drop of 0.3 V through the series resistances, the average field in the channel is 70 kV cm^{-1}. On such high fields, the stable state carrier velocity saturates, and this saturation velocity becomes an additional significant measure of carrier transport. Figure 8.5(c) displays plots of the electron velocity vs. the electric field for the conventional semiconductors, and simulated plots for huge-area graphene and a CNT (Pereira et al., 2009; Sano and Otsuji, 2009). For the nanotube and graphene, maximum carrier velocities of about 4×10^7 cm s^{-1} are expected, in comparison with 10^7 cm s^{-1} for silicon and 2×10^7 cm s^{-1} for GaAs. Furthermore, at high fields, the velocity in the nanotube and graphene does not drop as extremely as in the iii–v semiconductors. Inappropriately, there is currently no experimental data accessible on high-field transport in large-area graphene and GNRs. Though, other measurements propose high-field carrier velocities of numerous 10^7 cm s^{-1} in graphene. Therefore, regarding high-field transport, nanotubes, and graphene look to have a little advantage over conventional semiconductors (Sols et al., 2007; Ni et al., 2008).

Lastly, it is worth seeing that reported motilities for graphene devices are required to be interpreted wisely as there are numerous definitions for the MOSFET channel mobility, and they are tough to compare (Han et al., 2010). Moreover, the techniques utilized to measure mobility are only imprecisely described in certain papers. Most commonly, the field-effect mobility, μ_{FE}, is measured. Though, the impact of the source and drain series resistances must be eradicated from the measured characteristics to decide this quantity, and it is not constantly clear that this has been done (Cervantes-Sodi et al., 2008; Jiao et al., 2010).

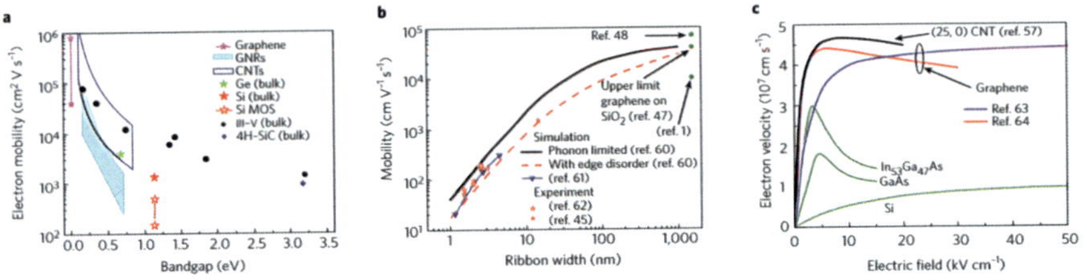

Figure 8.5. Carrier transport in graphene. (a) Electron mobility vs. bandgap in low electric fields for diverse materials; (b) carrier mobility vs. nanoribbon width at low electric fields from experiments and simulations; (c) electron drift velocity vs. electric field for usual semiconductors (Si, GaAs, In0.53Ga0.47As), large-area graphene, and a carbon nanotube.

Source: https://pubmed.ncbi.nlm.nih.gov/21715776/.

8.4. MOSFET GRAPHENE TRANSISTORS

A graphene MOS device was amongst the breakthrough results stated by the Manchester group in 2004. A 300 nm SiO_2 layer beneath the graphene worked as a back-gate dielectric and a doped silicon substrate proceeded as the back-gate (Figure 8.6(a)). Such back gate devices had been very beneficial for proof-of-concept drives; however, they suffer from unacceptably huge parasitic capacitances and cannot be incorporated with other components. Thus, practical graphene transistors required a top-gate. The initial graphene MOSFET with a top-gate was stated in 2007), demonstrating a significant milestone, and development had been very speedy since then (Figure 8.6(b)). Though graphene research is still in its initial stages, graphene MOSFETs could compete with devices that have been promoted from decades of investment and research (Chen et al., 2008; Raza and Kan, 2008).

Top-gated graphene MOSFETs had been prepared with exfoliated graphene, graphene grownup on metals like copper and nickel, Al_2O_3, HfO_2, SiO_2, and epitaxial graphene; had been utilized for the top-gate dielectric. The channels of these top gated graphene transistors had been made employing huge-area graphene, which does not have a bandgap, hence they have not been capable to switch off (Chen et al., 2009).

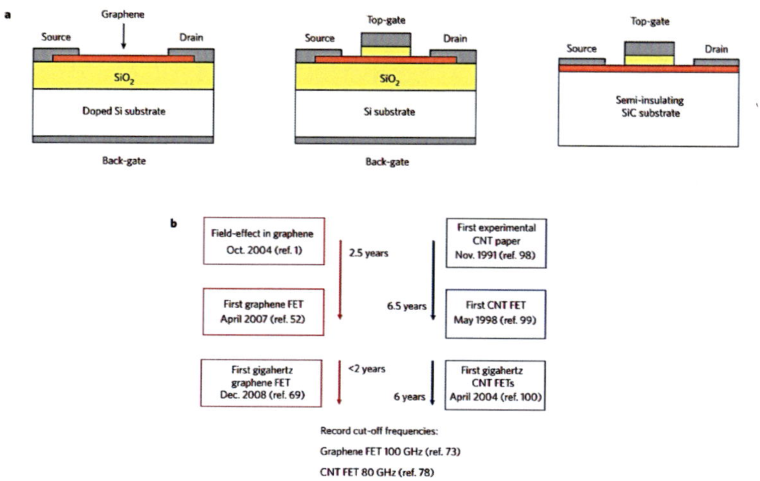

Figure 8.6. Evolution and structure of graphene MOSFETs. (a) Diagrams of different graphene MOSFET kinds: back-gated MOSFET (left); top-gated MOSFET with graphene grown on metal or a channel of exfoliated graphene and transferred to a SiO_2-covered Si wafer (middle); top-gated MOSFET with an epitaxial-graphene channel (right). (b) Development in graphene MOSFET development matched with the evolution of nanotube FETs.

Source: https://ieeexplore.ieee.org/document/6518205/.

Large-area-graphene transistors had a distinctive current-voltage transfer characteristic (Figure 8.7(a)). The type of carrier (holes and electrons) and carrier density in the channel are regulated by the potential differences amongst the gates (back-gate and/or top-gate) channel and the gates (back-gate and/or top-gate). Huge positive gate voltages encourage an electron gathering in the channel (n-type channel), and large negative gate voltages take to a p-type channel. This behavior provides growth to the two branches of the transfer features separated through the Dirac point (Figure 8.7(a)). The position of the Dirac point relies on numerous factors: the difference amongst the graphene and the work functions of the gate, the density and kind of the charges at the interfaces at the bottom and top of the channel (Figure 8.6), and any doping of the graphene. The on-off ratios described for MOSFET devices with large-area-graphene channels are to the extent of 2 to 20 (Morozov et al., 2008).

The output characteristics of numerous graphene MOSFETs either display a linear shape with only weak saturation or without any saturation, each of which is a drawback concerning device swiftness. Though, certain graphene MOSFETs had a rare type of saturation-like behavior that comprises a second linear region (Figure 8.7(b)). Our current understanding of the origin of this behavior is as follows. For little values of V_{DS}, the transistor functions in the linear region, and the whole channel is n-type (region I). As V_{DS} is enhanced, the drain current initiates to saturate till the inflection point at $V_{DS} = V_{DS,crit}$ is got (region II). At this stage, the potential situation at the drain end of the channel relates to the Dirac point. Once V_{DS} surpasses $V_{DS,crit}$, the conduction kind at the drain end of the channel shifts from n-type to p-type and the transistor goes into a second linear region (region III). At adequately large values of V_{DS}, the output characteristics for diverse gate voltages might cross, leading to 0 or even negative transconductance a highly adverse situation. This abnormal behavior is a result of these devices having gapless channels and does not happen in FETs with semiconducting channels (Schwierz, 2010).

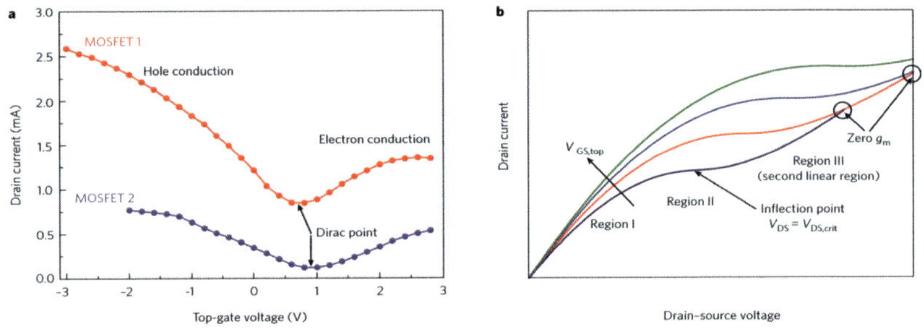

Figure 8.7. Direct-current behavior of graphene MOSFETs with a large-area-graphene channel. (a) Usual transfer characteristics for 2 MOSFETs with large-area-graphene channels. (b) Qualitative form of the output characteristics (drain current, ID, vs. drain-source voltage, VDS) of a MOSFET with an n-type large-area-graphene channel, for diverse values of the VGS, top-gate voltage, top.

Source: https://ieeexplore.ieee.org/document/5061601.

REFERENCES

1. Aberg, I., & Hoyt, J. L., (2005). Hole transport in UTB MOSFETs in strained-Si directly on insulator with strained-Si thickness less than 5 nm. *IEEE Electron Device Letters, 26*(9), 661–663.
2. Berger, C., Song, Z., Li, X., Wu, X., Brown, N., Naud, C., & Conrad, E. H., (2006). Electronic confinement and coherence in patterned epitaxial graphene. *Science, 312*(5777), 1191–1196.
3. Boehm, H. P., Clauss, A., Fischer, G. O., & Hofmann, U., (1962). Dünnste kohlenstoff-folien. *Zeitschrift Für Naturforschung B,17*(3), 150–153.
4. Castro, E. V., Novoselov, K. S., Morozov, S. V., Peres, N. M. R., Dos, S. J. L., Nilsson, J., & Neto, A. C., (2007). Biased bilayer graphene: Semiconductor with a gap tunable by the electric field effect. *Physical Review Letters, 99*(21), 216802.
5. Cervantes-Sodi, F., Csányi, G., Piscanec, S., & Ferrari, A. C., (2008). Edge-functionalized and substitutionally doped graphene nanoribbons: Electronic and spin properties. *Physical Review B,77*(16), 165427.
6. Chau, R., Doyle, B., Datta, S., Kavalieros, J., & Zhang, K., (2007). Integrated nanoelectronics for the future. *Nature Materials, 6*(11), 810–812.
7. Chen, F., Xia, J., Ferry, D. K., & Tao, N., (2009). Dielectric screening enhanced performance in graphene FET. *Nano Letters, 9*(7), 2571–2574.
8. Chen, J. H., Jang, C., Xiao, S., Ishigami, M., & Fuhrer, M. S., (2008). Intrinsic and extrinsic performance limits of graphene devices on SiO_2. *Nature Nanotechnology, 3*(4), 206–209.
9. Chen, Z., Lin, Y. M., Rooks, M. J., & Avouris, P., (2007). Graphene nano-ribbon electronics. *Physica E: Low-Dimensional Systems and Nanostructures, 40*(2), 228–232.
10. Evaldsson, M., Zozoulenko, I. V., Xu, H., & Heinzel, T., (2008). Edge-disorder-induced Anderson localization and conduction gap in graphene nanoribbons. *Physical Review B, 78*(16), 161407.
11. Frank, D. J., Taur, Y., & Wong, H. S., (1998). Generalized scale length for two-dimensional effects in MOSFETs. *IEEE Electron Device Letters, 19*(10), 385–387.
12. Gava, P., Lazzeri, M., Saitta, A. M., & Mauri, F., (2009). Ab initio study of gap opening and screening effects in gated bilayer graphene.

Physical Review B,79(16), 165431.

13. Geim, A. K., & Novoselov, K. S., (2010). The rise of graphene. In: *Nanoscience and Technology: A Collection of Reviews from Nature Journals* (Vol. 1, pp. 11–19).
14. Geim, A. K., (2009). Graphene: Status and prospects. *Science, 324*(5934), 1530–1534.
15. Grau, G., (2011). Road-mapping as a tool for new product introduction (NPI) illustrated by the example of the semiconductor industry. *Interdisciplinary Management Research, 7*, 1–22.
16. Han, M. Y., Brant, J. C., & Kim, P., (2010). Electron transport in disordered graphene nanoribbons. *Physical Review Letters, 104*(5), 056801.
17. Han, M. Y., Özyilmaz, B., Zhang, Y., & Kim, P., (2007). Energy band-gap engineering of graphene nanoribbons. *Physical Review Letters, 98*(20), 206805.
18. Hughes, B., & Tasker, P. J., (1989). Bias dependence of the MODFET intrinsic model elements values at microwave frequencies. *IEEE Transactions on Electron Devices, 36*(10), 2267–2273.
19. Jiao, L., Wang, X., Diankov, G., Wang, H., & Dai, H., (2010). Facile synthesis of high-quality graphene nanoribbons. *Nature Nanotechnology, 5*(5), 321–325.
20. Kedzierski, J., Hsu, P. L., Healey, P., Wyatt, P. W., Keast, C. L., Sprinkle, M., & De Heer, W. A., (2008). Epitaxial graphene transistors on SiC substrates. *IEEE Transactions on Electron Devices, 55*(8), 2078–2085.
21. Kim, K. S., Zhao, Y., Jang, H., Lee, S. Y., Kim, J. M., Kim, K. S., & Hong, B. H., (2009). Large-scale pattern growth of graphene films for stretchable transparent electrodes. *Nature, 457*(7230), 706–710.
22. Kim, S., Ihm, J., Choi, H. J., & Son, Y. W., (2008). Origin of anomalous electronic structures of epitaxial graphene on silicon carbide. *Physical Review Letters, 100*(17), 176802.
23. Li, X., Wang, X., Zhang, L., Lee, S., & Dai, H., (2008). Chemically derived, ultrasmooth graphene nanoribbon semiconductors. *Science, 319*(5867), 1229–1232.
24. May, J. W., (1969). Platinum surface LEED rings. *SurSc, 17*(1), 267–270.

25. Morozov, S. V., Novoselov, K. S., Katsnelson, M. I., Schedin, F., Elias, D. C., Jaszczak, J. A., & Geim, A. K., (2008). Giant intrinsic carrier mobilities in graphene and its bilayer. *Physical Review Letters, 100*(1), 016602.
26. Neto, A. C., Guinea, F., Peres, N. M., Novoselov, K. S., & Geim, A. K., (2009). The electronic properties of graphene. *Reviews of Modern Physics, 81*(1), 109.
27. Nguyen, L. D., Tasker, P. J., Radulescu, D. C., & Eastman, L. F., (1989). Characterization of ultra-high-speed pseudomorphic AlGaAs/InGaAs (on GaAs) MODFETs. *IEEE Transactions on Electron Devices, 36*(10), 2243–2248.
28. Ni, Z. H., Yu, T., Lu, Y. H., Wang, Y. Y., Feng, Y. P., & Shen, Z. X., (2008). Uniaxial strain on graphene: Raman spectroscopy study and band-gap opening. *ACS Nano, 2*(11), 2301–2305.
29. Ning, T. H., & Taur, Y., (1998). *Fundamentals of Modern VLSI Devices* (Vol. 1, pp. 1–20).
30. Novoselov, K. S., Geim, A. K., Morozov, S. V., Jiang, D., Zhang, Y., Dubonos, S. V., & Firsov, A. A., (2004). Electric field effect in atomically thin carbon films. *Science, 306*(5696), 666–669.
31. Ohta, T., Bostwick, A., Seyller, T., Horn, K., & Rotenberg, E., (2006). Controlling the electronic structure of bilayer graphene. *Science, 313*(5789), 951–954.
32. Peng, X., & Ahuja, R., (2008). Symmetry breaking induced bandgap in epitaxial graphene layers on SiC. *Nano Letters, 8*(12), 4464–4468.
33. Pereira, V. M., Neto, A. C., & Peres, N. M. R., (2009). Tight-binding approach to uniaxial strain in graphene. *Physical Review B,80*(4), 045401.
34. Raza, H., & Kan, E. C., (2008). Armchair graphene nanoribbons: Electronic structure and electric-field modulation. *Physical Review B, 77*(24), 245434.
35. Reina, A., Jia, X., Ho, J., Nezich, D., Son, H., Bulovic, V., & Kong, J., (2009). Large area, few-layer graphene films on arbitrary substrates by chemical vapor deposition. *Nano Letters, 9*(1), 30–35.
36. Rotenberg, E., Bostwick, A., Ohta, T., McChesney, J. L., Seyller, T., & Horn, K., (2008). Origin of the energy bandgap in epitaxial graphene. *Nature Materials, 7*(4), 258–259.

37. Sano, E., & Otsuji, T., (2009). Theoretical evaluation of channel structure in graphene field-effect transistors. *Japanese Journal of Applied Physics, 48*(4R), 041202.
38. Schwierz, F., & Liou, J. J., (2003). *Modern Microwave Transistors: Theory, Design, and Performance* (Vol. 1, pp. 1–32). Wiley-Interscience.
39. Schwierz, F., & Liou, J. J., (2007). RF transistors: Recent developments and roadmap toward terahertz applications. *Solid-State Electronics, 51*(8), 1079–1091.
40. Schwierz, F., (2010). Graphene transistors. *Nature Nanotechnology, 5*(7), 487.
41. Schwierz, F., Wong, H., & Liou, J. J., (2010). *Nanometer CMOS* (Vol. 1, pp. 1–22). Pan Stanford Publishing.
42. Sols, F., Guinea, F., & Neto, A. C., (2007). Coulomb blockade in graphene nanoribbons. *Physical Review Letters, 99*(16), 166803.
43. Thompson, S. E., Chau, R. S., Ghani, T., Mistry, K., Tyagi, S., & Bohr, M. T., (2005). In search of "forever," continued transistor scaling one new material at a time. *IEEE Transactions on Semiconductor Manufacturing, 18*(1), 26–36.
44. Uyemura, J. P., (1999). *CMOS Logic Circuit Design* (Vol. 1, pp. 1–32). Springer Science & Business Media.
45. Van, B. A. J., Crombeen, J. E., & Van, T. A., (1975). LEED and Auger electron observations of the SiC (0001) surface. *Surface Science, 48*(2), 463–472.
46. Yang, L., Park, C. H., Son, Y. W., Cohen, M. L., & Louie, S. G., (2007). Quasiparticle energies and band gaps in graphene nanoribbons. *Physical Review Letters, 99*(18), 186801.
47. Young, I. A., Mohammed, E., Liao, J. T., Kern, A. M., Palermo, S., Block, B. A., & Chang, P. L., (2009). Optical I/O technology for tera-scale computing. *IEEE Journal of Solid-State Circuits, 45*(1), 235–248.
48. Zhang, Y., Tang, T. T., Girit, C., Hao, Z., Martin, M. C., Zettl, A., & Wang, F., (2009). Direct observation of a widely tunable bandgap in bilayer graphene. *Nature, 459*(7248), 820–823.
49. Zhou, S. Y., Gweon, G. H., Fedorov, A. V., First, P. D., De Heer, W. A., Lee, D. H., & Lanzara, A., (2007). Substrate-induced bandgap opening in epitaxial graphene. *Nature Materials, 6*(10), 770–775.

Index

D

E

F

G

H

I

L

M

N

O

P

Q

R

S

T

U

W

X

Z